Annals of Mathematics Studies

Number 135

Complex Dynamics and Renormalization

by

Curtis T. McMullen

PRINCETON UNIVERSITY PRESS

———

PRINCETON, NEW JERSEY

1994

The Annals of Mathematics Studies are edited by
Luis A. Caffarelli, John N. Mather, and Elias M. Stein

Princeton University Press books are printed on acid-free paper and meet the
guidelines for permanence and durability of the Committee on Production
Guidelines for Book Longevity of the Council on Library Resources

Printed in the United States of America

10 9 8 7 6 5 4 3 2 1

Library of Congress Cataloging-in-Publication Data

McMullen, Curtis T.
Complex dynamics and renormalization / Curtis T. McMullen.
p. cm. (Annals of mathematics studies : no. 135)
Includes bibliographical references and index.
ISBN 0-691-02982-2 (cloth)   ISBN 0-691-02981-4 (pbk.)
1. Renormalization. 2. Polynomials. 3. Dynamics. 4. Mathematical physics.
I. Title. II. Series.
QC20.7.R43M36   1994
530.1'43'0151 dc20   94-29390

The publisher would like to acknowledge the author of this volume for providing
the camera-ready copy from which this book was printed

Contents

Complex Dynamics and Renormalization

Chapter 1

Introduction

1.1 Complex dynamics

This work presents a study of renormalization of quadratic polynomials and a rapid introduction to techniques in complex dynamics.

Around 1920 Fatou and Julia initiated the theory of iterated rational maps

$$f : \widehat{\mathbb{C}} \to \widehat{\mathbb{C}}$$

on the Riemann sphere. More recently methods of geometric function theory, quasiconformal mappings and hyperbolic geometry have contributed to the depth and scope of research in the field. The intricate structure of the family of quadratic polynomials was revealed by work of Douady and Hubbard [DH1], [Dou1]; analogies between rational maps and Kleinian groups surfaced with Sullivan's proof of the no wandering domains theorem [Sul3] and continue to inform both subjects [Mc2].

It can be a subtle problem to understand a high iterate of a rational map f of degree $d > 1$. There is tension between expanding features of f — such as the fact that its degree tends to infinity under iteration — and contracting features, such as the presence of critical points. The best understood maps are those for which the critical points tend to attracting cycles. For such a map, the tension is resolved by the concentration of expansion in the *Julia set* or chaotic locus of the map, and the presence of contraction on the rest of the sphere.

1

The central goal of this work is to understand a high iterate of a quadratic polynomial. The special case we consider is that of an *infinitely renormalizable* polynomial $f(z) = z^2 + c$.

For such a polynomial, the expanding and contracting properties lie in a delicate balance; for example, the critical point $z = 0$ belongs to the Julia set and its forward orbit is recurrent. Moreover high iterates of f can be *renormalized* or rescaled to yield new dynamical systems of the same general shape as the original map f.

This repetition of form at infinitely many scales provides the basic framework for our study. Under additional geometric hypotheses, we will show that the renormalized dynamical systems range in a compact family. Compactness is established by combining universal estimates for the hyperbolic geometry of surfaces with distortion theorems for holomorphic maps.

With this information in hand, we establish quasiconformal rigidity of the original polynomial f. Rigidity of f supports conjectures about the behavior of a generic complex dynamical system, as described in the next section.

The course of the main argument entails many facets of complex dynamics. Thus the sequel includes a brief exposition of topics including:

- The Poincaré metric, the modulus of an annulus, and distortion theorems for univalent maps (§2);

- The collar theorem and related aspects of hyperbolic surfaces (§2.9 and §2.10);

- Dynamics of rational maps and hyperbolicity (§3.1 and §3.4);

- Ergodic theory of rational maps, and the role of the postcritical set as a measure-theoretic attractor (§3);

- Invariant line fields, holomorphic motions and stability in families of rational maps (§3 and §4);

- The Mandelbrot set (§4);

- Polynomial-like maps and proper maps (§5);

- Riemann mappings and external rays (§5 and §6);

- Renormalization (§7);

- The Yoccoz puzzle (§8);

- Real methods and Sullivan's *a priori* bounds (§11);

- Orbifolds (Appendix A); and

- Thurston's characterization of critically finite rational maps (Appendix B).

1.2 Central conjectures

We now summarize the main problems which motivate our work.

Let $f : \widehat{\mathbb{C}} \to \widehat{\mathbb{C}}$ be a rational map of the Riemann sphere to itself of degree $d > 1$. The map f is *hyperbolic* if its critical points tend to attracting periodic cycles under iteration. Within all rational maps, the hyperbolic ones are among the best behaved; for example, when f is hyperbolic there is a finite set $A \subset \widehat{\mathbb{C}}$ which attracts all points in an open, full-measure subset of the sphere (see §3.4).

One of the central problems in conformal dynamics is the following:

Conjecture 1.1 (Density of hyperbolicity) *The set of hyperbolic rational maps is open and dense in the space* Rat_d *of all rational maps of degree* d.

Openness of hyperbolic maps is known, but density is not. In some form this conjecture goes back to Fatou (see §4.1).

Much study has been devoted to special families of rational maps, particularly quadratic polynomials. Every quadratic polynomial f is conjugate to one of the form $f_c(z) = z^2 + c$ for a unique $c \in \mathbb{C}$. Even this simple family of rational maps exhibits a full spectrum of dynamical behavior, reflecting many of the difficulties of the general case. Still unresolved is:

Conjecture 1.2 *The set of* c *for which* $z^2 + c$ *is hyperbolic forms an open dense subset of the complex plane.*

The *Mandelbrot set* M is the set of c such that under iteration, $f_c^n(0)$ does not tend to infinity; here $z = 0$ is the unique critical point of f_c in $\mathbb{C}$. A component U of the interior of M is *hyperbolic* if f_c is hyperbolic for some c in U. It is known that the maps f_c enjoy a type of structural stability as c varies in any component of $\mathbb{C} - \partial M$; in particular, if U is hyperbolic, f_c is hyperbolic for every c in U (see §4). It is clear that f_c is hyperbolic when c is not in M, because the critical point tends to the superattracting fixed point at infinity. Thus an equivalent formulation of Conjecture 1.2 is:

Conjecture 1.3 *Every component of the interior of the Mandelbrot set is hyperbolic.*

An approach to these conjectures is developed in [MSS] and [McS], using quasiconformal mappings. This approach has the advantage of shifting the focus from a family of maps to the dynamics of a single map, and leads to the following:

Conjecture 1.4 (No invariant line fields) *A rational map f carries no invariant line field on its Julia set, except when f is double covered by an integral torus endomorphism.*

Conjecture 1.4 implies all the preceding conjectures [McS]. This conjecture is explained in more detail in §3.5; see also [Mc3].

The rational maps which are covered by integral torus endomorphisms form a small set of exceptional cases. For quadratic polynomials, Conjecture 1.4 specializes to:

Conjecture 1.5 *A quadratic polynomial carries no invariant line field on its Julia set.*

The *Julia set* J of a polynomial f is the boundary of the set of points which tend to infinity under iteration. A *line field* on J is the assignment of a real line through the origin in the tangent space to z for each z in a positive measure subset E of J, so that the slope is a measurable function of z. A line field is *invariant* if $f^{-1}(E) = E$, and if f' transforms the line at z to the line at $f(z)$.

Conjecture 1.5 is *equivalent* to Conjectures 1.2 and 1.3 (see §4).

Recent progress towards these conjectures includes:

Theorem 1.6 (Yoccoz) *A quadratic polynomial which carries an invariant line field on its Julia set is infinitely renormalizable.*

See §8. Here a quadratic polynomial is *infinitely renormalizable* if there are infinitely many $n > 1$ such that f^n restricts to a quadratic-like map with connected Julia set; see §7. For instance, the much-studied Feigenbaum example is an infinitely renormalizable polynomial (see §7.4).

This work addresses the infinitely renormalizable case. Our main result is:

Theorem 1.7 (Robust rigidity) *A robust infinitely renormalizable quadratic polynomial f carries no invariant line field on its Julia set.*

See §10. Roughly speaking, a quadratic polynomial is *robust* if it admits infinitely many renormalizations with definite space around the small postcritical sets (see §9).

To establish this result, we will show that suitable renormalizations of a robust quadratic polynomial range through a compact set of proper mappings. One may compare our proof of the absence of invariant line fields to a fundamental result of Sullivan, which states that the limit set of a finitely generated Kleinian group carries no invariant line field [Sul1]. The compactness of renormalizations plays a role something like the finite-dimensionality of the ambient Lie group for a Kleinian group.

It can be shown that every infinitely renormalizable real quadratic polynomial is robust (§11). When combined with the result of Yoccoz, we obtain:

Corollary 1.8 *The Julia set of a real quadratic polynomial carries no invariant line field.*

From the λ-lemma of [MSS], one obtains:

Corollary 1.9 *Every component of the interior of the Mandelbrot set meeting the real axis is hyperbolic.*

These corollaries are deduced in §11.

1.3 Summary of contents

We begin in §2 with a resume of results from hyperbolic geometry, geometric function theory and measure theory. Then we introduce the theory of iterated rational maps, and study their measurable dynamics in §3.

Here one may see the first instance of a general philosophy:

> *Expanding dynamics promotes a measurable line field to a holomorphic line field.*

This philosophy has precursors in [Sul1] and classical arguments in ergodic theory.

In §4 we discuss holomorphic motions and structural stability in general families of rational maps. Then we specialize to the Mandelbrot set, and explain the equivalence of Conjectures 1.2 and 1.5.

In §5, we develop compactness results to apply the expansion philosophy in the context of renormalization. We also introduce the *polynomial-like maps* of Douady and Hubbard, which play a fundamental role in renormalization.

In §6, we turn to polynomials and describe the use of external rays in the study of their combinatorics.

With this background in place, the theory of renormalization is developed in §7. New types of renormalization, unrelated to "tuning", were discovered in the course of this development; examples are presented in §7.4.

§8 describes infinitely renormalizable quadratic polynomials. Included is an exposition of the Yoccoz puzzle, a Markov partition for the dynamics of a quadratic polynomial. Theorem 1.6 is discussed along with work of Lyubich and Shishikura.

In §9 we define *robust* quadratic polynomials, and prove their postcritical sets have measure zero. This assertion is essential for applying the expansion philosophy, because we only obtain expansion in the complement of the postcritical set.

§10 gives the proof of Theorem 1.7(Robust rigidity). The proof is broken down into two cases. In the first case, the postcritical set falls into far-separated blocks at infinitely many levels of renormalization. Using this separation, we extract a polynomial-like map g as a limit of infinitely many renormalizations of a quadratic polynomial f. If

f carries a measurable invariant line field, the expansion philosophy leads to a well-behaved holomorphic invariant line field for g, which can easily be shown not to exist.

In the second case, the blocks of the postcritical set are not well-separated. Then we carry out a parallel argument *without* attempting to produce a polynomial-like limit of renormalization. More flexible limits of renormalization still suffice to give rigidity of the original quadratic map. The limit constructed in this case is a proper map of degree two $g : X \to Y$, between disks X and Y, with the critical point of g in $X \cap Y$. (We do not require $X \subset Y$.) The dynamics of g is sufficiently nonlinear to again rule out the existence of a measurable line field for the original map f.

A useful tool in our study of complex renormalization is the following result from hyperbolic geometry. Let X be a surface of finite area with one cusp and geodesic boundary. Suppose each boundary component has length at least L. Then two boundary components are within distance $D(L)$ of each other (Theorem 2.24). In the complex setting, this result will substitute for the "shortest interval argument" in one dimensional real dynamics.

In §11 we recapitulate and extend arguments of Sullivan to show that an infinitely renormalizable *real* quadratic polynomial $z^2 + c$ is robust. This gives Corollaries 1.8 and 1.9 above.

Appendix A provides background on orbifolds, including the uniformization theorem.

Appendix B further develops the foundations of renormalization, by introducing the notion of a *quotient map* between two branched covers of the sphere. We prove any critically finite quotient of a rational map is again rational. This result can be thought of as a 'closing lemma' for rational maps, although we do not show that the critically finite quotient map is near to the original one. With this result one may construct an infinite sequence g_n of quadratic polynomials canonically associated to an infinitely renormalizable quadratic polynomial. Conjecturally, $g_n \to f$; this conjecture also implies the density of hyperbolic dynamics in the quadratic family.

Related literature. Basic material on iterated rational maps can be found in [Fatou1], [Fatou2], [Fatou3], [Julia], [Bro], [Dou1], [McS], [Bl], [Mil2], [EL] [Bea2] and [CG]. A survey of the conjectures in conformal dynamics which motivate this work appears in [Mc3].

Renormalization is a broad topic, many aspects of which we do not touch on here. An exposition of results of Branner, Hubbard and Yoccoz and their relation to complex renormalization appears in [Mil3]. See [Cvi] for a collection of papers on the discovery and development of renormalization. Fundamental results on compactness and convergence of renormalization for real quadratic maps appear in [Sul4]; see also [MeSt] for a treatment of Sullivan's results. Another approach to convergence of renormalization, via rigidity of towers, appears in [Mc2] and [Mc4]. The relation of renormalization to self-similarity in the Mandelbrot set is studied in [Mil1].

The conjectures that we study here are a field of active research; in particular, Lyubich and Świątek have independently made deep contributions towards the density of expanding dynamics in the quadratic family [Lyu3], [Sw].

A preliminary version of this manuscript was written in summer of 1992. This research was partially supported by the NSF, IHES and the Sloan Foundation.

Chapter 2

Background in conformal geometry

This chapter begins with standard results in geometric function theory, quasiconformal mappings, hyperbolic geometry and measure theory that will be used in the sequel. These results describe the approximate geometry of annuli, univalent maps, measurable sets and hyperbolic surfaces. The reader may wish to concentrate on the statements rather than the proofs, which are sometimes technical.

We also include geometric theorems needed in the sequel that can be stated without reference to dynamics: a measure zero criterion (§2.8), the collar theorem (§2.9), a complex version of the shortest interval argument (§2.10) and bounds on holomorphic contraction (§2.11).

Notation. The Riemann sphere and the punctured plane will be denoted by:

$$\begin{aligned} \widehat{\mathbb{C}} &= \mathbb{C} \cup \{\infty\} \quad \text{and} \\ \mathbb{C}^* &= \mathbb{C} - \{0\}; \end{aligned}$$

the upper halfplane, unit disk and punctured disk by:

$$\begin{aligned} \mathbb{H} &= \{z \; : \; \mathrm{Im}(z) > 0\}, \\ \Delta &= \{z \; : \; |z| < 1\} \quad \text{and} \\ \Delta^* &= \Delta - \{0\}; \end{aligned}$$

9

and a family of annuli centered at zero by

$$A(R) \ = \ \{z \ : \ 1 < |z| < R\}.$$

The disk of radius r will be denoted by

$$\Delta(r) \ = \ \{z \ : \ |z| < r\},$$

and the unit circle by

$$S^1 \ = \ \{z \ : \ |z| = 1\}.$$

A map of pairs $f : (A, A') \to (B, B')$ means a map $f : A \to B$ such that $f(A') \subset B'$. The restriction of a mapping f to a subset U with $f(U) \subset V$ will be denoted simply by $f : U \to V$.

$O(x)$ denotes a quantity whose absolute value is bounded by Cx for some unspecified universal constant C; $q \asymp x$ means $cx < q < Cx$, again for unspecified $c, C > 0$.

Bounds of the form $A < C(B)$ mean A is bounded by a quantity which depends only on B. Different occurrences of $C(B)$ are meant to be independent.

2.1   The modulus of an annulus

Any Riemann surface A with $\pi_1(A) \cong \mathbb{Z}$ is isomorphic to $\mathbb{C}^*$, Δ^* or the standard annulus $A(R)$ for some $R \in (1, \infty)$. In case A is isomorphic to $A(R)$, the *modulus* of A is defined by

$$\mathrm{mod}(A) \ = \ \frac{\log R}{2\pi}.$$

Thus A is conformally isomorphic to a right cylinder of circumference one and height $\mathrm{mod}(A)$. By convention $\mathrm{mod}(A) = \infty$ in the other two cases.

An annulus $B \subset \mathbb{C}$ is *round* if it is bounded by concentric Euclidean circles (so B has the form $\{z \ : \ r < |z - c| < s\}$).

Theorem 2.1 (Round annulus) *Any annulus $A \subset \mathbb{C}$ of sufficiently large modulus contains an essential round annulus B with $\mathrm{mod}(A) = \mathrm{mod}(B) + O(1)$.*

Here *essential* means $\pi_1(B)$ injects into $\pi_1(A)$, i.e. B separates the boundary components of A.

Proof. We may assume $\hat{\mathbb{C}} - A$ consists of two components C and D, where $0 \in C$ and $\infty \in D$. Let $z_1 \in C$ maximize $|z|$ over C, and let $z_2 \in D$ minimize $|z|$ over D. By Teichmüller's module theorem [LV, §II.1.3],

$$\text{mod}(A) \le \frac{1}{\pi} \mu \left(\sqrt{\frac{|z_1|}{|z_1| + |z_2|}} \right)$$

where $\mu(r)$ is a positive decreasing function of r.[1] Thus $|z_1| < |z_2|$ if $\text{mod}(A)$ is sufficiently large, in which case A contains a round annulus $B = \{ z \; : \; |z_1| < |z| < |z_2| \}$. Moreover, once $|z_1| < |z_2|$ we have

$$\text{mod}(A) \le \frac{1}{\pi} \mu \left(\sqrt{\frac{|z_1|}{2|z_2|}} \right) \le \text{mod}(B) + \frac{5 \log 2}{2\pi}$$

by the inequality $\mu(r) < \log(4/r)$ [LV, eq. (2.10) in §II.2.3]. ∎

An alternative proof can be based on the following fact: any sequence of univalent maps $f_n : \{ z \; : \; 1/R_n < |z| < R_n \} \to \mathbb{C}^*$, with $f_n(1) = 1$ and with the image of f separating 0 from ∞, converges to the identity as $R_n \to \infty$.

2.2   The hyperbolic metric

A Riemann surface is *hyperbolic* if its universal cover is isomorphic to the upper halfplane $\mathbb{H}$. The *hyperbolic metric* or *Poincaré metric* on such a Riemann surface is the unique complete conformal metric of constant curvature -1.

By the Schwarz lemma [Ah2, §1-2] one knows:

Theorem 2.2 *A holomorphic map* $f : X \to Y$ *between hyperbolic Riemann surfaces does not increase the Poincaré metric, and* f *is a local isometry if and only if* f *is a covering map.*

---

[1] In [LV] the modulus of $A(R)$ is defined to be $\log(R)$ rather than $\log(R)/2\pi$.

The Poincaré metric is defined on any region $U \subset \hat{\mathbb{C}}$ provided $|\hat{\mathbb{C}} - U| > 2$. If U is not connected, we define its Poincaré metric component by component.

The hyperbolic metric on the upper halfplane $\mathbb{H}$ is given by:

$$\rho = \frac{|dz|}{\mathrm{Im}(z)};$$

on the unit disk Δ, by:

$$\rho = \frac{2|dz|}{1 - |z|^2};$$

on the punctured disk Δ^*, by:

$$\rho = \frac{|dz|}{|z| \log(1/|z|)};$$

and on the annulus $A(R)$ by:

$$\rho = \frac{\pi/\log R}{\sin(\pi \log|z|/\log R)} \frac{|dz|}{|z|}.$$

The last two formulas can be verified using the covering maps $z \mapsto \exp(iz)$ from $\mathbb{H}$ to Δ^* and $z \mapsto z^{\log R/\pi i}$ from $\mathbb{H}$ to $A(R)$.

The *core curve* γ of an annulus X of finite modulus is its unique closed geodesic. The hyperbolic length of γ is related to the modulus by

$$\mathrm{length}(\gamma) = \frac{\pi}{\mathrm{mod}(X)},$$

as can be checked by considering the circle $|z| = \sqrt{R}$ that forms the core curve of $A(R)$.

It is useful to keep in mind an approximate picture for the Poincaré metric on an arbitrary region U in the plane. Such a picture is provided by a theorem of Beardon and Pommerenke [BP, Theorem 1], which we formulate as follows.

Let $d(z, \partial U)$ be the Euclidean distance from z to the boundary of U. Let $\mathrm{mod}(z, U)$ denote the maximum modulus of an essential round annulus in U whose core curve passes through z. If no such annulus exists, set $\mathrm{mod}(z, U) = 0$.

Theorem 2.3 (Poincaré metric on a plane region) *For any hyperbolic region U in the plane, the Poincaré metric ρ is comparable to*

$$\rho' = \frac{|dz|}{d(z, \partial U)(1 + \text{mod}(z, U))}.$$

That is, $1/C < (\rho/\rho') < C$ for some universal constant $C > 0$.

This theorem can also be derived from the thick-thin decomposition for hyperbolic manifolds and Theorem 2.1 above.

2.3 Metric aspects of annuli

Let V be a Riemann surface which is topologically a disk, and let $E \subset V$ have compact closure. It is convenient to have a measurement of the amount of space around E in V. For this purpose we define

$$\text{mod}(E, V) = \sup\, \{\text{mod}(A)\ :\ A \subset V \text{ is an annulus enclosing } E\}.$$

(This means E should lie in the compact component of $V - A$.) Note that $\text{mod}(E, V) = \infty$ if V is isomorphic to $\mathbb{C}$ or if E is a single point.

Now suppose V is hyperbolic, and let $\text{diam}(E)$ denote diameter of E in the hyperbolic metric on V.

Theorem 2.4 *The hyperbolic diameter and modulus of E are inversely related:*

$$\text{diam}(E) \to 0 \quad \Longleftrightarrow \quad \text{mod}(E, V) \to \infty$$

and

$$\text{diam}(E) \to \infty \quad \Longleftrightarrow \quad \text{mod}(E, V) \to 0.$$

More precisely,

$$\text{diam}(E) \asymp \exp(-\text{mod}(E, V))$$

when either side is small, while

$$\frac{C_1}{\text{diam}(E)} \geq \text{mod}(E, V) \geq C_2 \exp(-\text{diam}(E))$$

when the diameter is large.

Proof. The first estimate follows from existence of a round annulus as guaranteed by Theorem 2.1. The second follows using estimates for the Grötzsch modulus [LV, §II.2].

∎

The relation of modulus to hyperbolic diameter is necessarily imprecise when the diameter is large. For example, for $r < 1$ the sets $E_1 = [-r, r]$ and $E_2 = \Delta(r)$ have the same hyperbolic diameter d in the unit disk, but for r near 1, $\mathrm{mod}(E_1, \Delta) \asymp 1/d$ while $\mathrm{mod}(E_2, \Delta) \asymp e^{-d}$.

The next result controls the Euclidean geometry of an annulus of definite modulus.

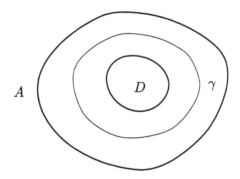

Figure 2.1. Core geodesic of an annulus.

Theorem 2.5 *Let $A \subset \mathbb{C}$ be an annulus with core curve γ and with modulus $\mathrm{mod}(A) > m > 0$. Let D the bounded component of $\mathbb{C} - A$. Then in the Euclidean metric,*

$$d(D, \gamma) > C(m)\,\mathrm{diam}(\gamma)$$

where $C(m) > 0$.

See Figure 2.1.

Proof. Since A contains an annulus of modulus m with the same core curve, it suffices to prove the theorem when $\mathrm{mod}(A) = m$.

Let x be a point in D. We may normalize coordinates on $\mathbb{C}$ so that $x = 0$ and $\text{diam}(\gamma) = 1$. Let $R = \exp(2\pi m)$. Then A does not contain the circle $|z| = R$, because otherwise $\text{mod}(A) > m$. By further normalizing with a rotation we can assume $A \subset \widehat{\mathbb{C}} - \{0, R, \infty\}$. The hyperbolic length of γ on A is π/m (see §2.2), so by the Schwarz lemma its length is less than π/m in the Poincaré metric on $\widehat{\mathbb{C}} - \{0, R, \infty\}$. Since the Euclidean diameter of γ is one and the hyperbolic metric on $\widehat{\mathbb{C}} - \{0, R, \infty\}$ is complete, we have $d(\gamma, 0) > C(m) > 0$. Equivalently, $d(x, \gamma) > C(m)\,\text{diam}(\gamma)$. Since x was an arbitrary point in D, the theorem follows.

■

2.4 Univalent maps

A *univalent map* f is an injective holomorphic map. The Koebe distortion theorems make precise the fact that a univalent map has bounded geometry; we summarize this principle as follows:

Theorem 2.6 (Koebe distortion) *The space of univalent maps*

$$f : \Delta \to \widehat{\mathbb{C}}$$

is compact up to post-composition with automorphisms of $\widehat{\mathbb{C}}$.

This means any sequence of univalent maps contains a subsequence $f_n : \Delta \to \widehat{\mathbb{C}}$ such that $M_n \circ f_n$ converges to a univalent map f, uniformly on compact subsets of Δ, for some sequence of Möbius transformations $M_n : \widehat{\mathbb{C}} \to \widehat{\mathbb{C}}$.

An equivalent and more classical formulation is the following (cf. [Ah2]):

Theorem 2.7 *The space S of univalent functions*

$$f : \Delta \to \mathbb{C},$$

normalized by $f(0) = 0$ and $f'(0) = 1$, is compact in the topology of uniform convergence on compact sets. In particular, for $r < 1$ and x, y in $\Delta(r)$ we have

$$\frac{1}{C(r)} \leq \frac{|f(x) - f(y)|}{|x - y|} \leq C(r)$$

and

$$\frac{1}{C(r)} \leq |f'(x)| \leq C(r)$$

for all f in S, where $C(r) \to 1$ as $r \to 0$.

An example of a normalized univalent map is shown in Figure 2.2.

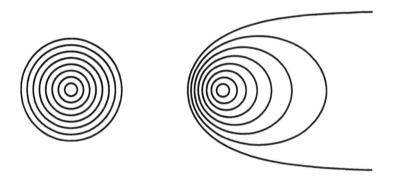

Figure 2.2. The unit disk and its image under $f(z) = \log(1 + z)$.

Corollary 2.8 *Let (X, x) be a (connected) Riemann surface with basepoint x, and let $\mathcal{F}$ denote the space of all univalent maps $f : (X, x) \to (\mathbb{C}, 0)$ such that $\|f'(x)\| = 1$ for some fixed metric on X and for the Euclidean metric on $\mathbb{C}$.*

Then $\mathcal{F}$ is compact in the topology of uniform convergence on compact sets.

Proof. Cover X with charts isomorphic to Δ. The restriction of $\mathcal{F}$ to a chart containing x is compact by the preceding results. When two charts U_1 and U_2 overlap at a point y, compactness of $\mathcal{F}|U_1$ implies upper and lower bounds on $\|f'(y)\|$, which implies compactness of $\mathcal{F}|U_2$. Connectedness of X completes the proof.

■

The Koebe principle also controls univalent maps defined on disks which are not round. In this case one obtains bounded geometry after discarding an annulus of definite modulus.

Theorem 2.9 *Let $D \subset U \subset \mathbb{C}$ be disks with $\mathrm{mod}(D, U) > m > 0$. Let $f : U \to \mathbb{C}$ be a univalent map. Then there is a constant $C(m)$ such that for any x, y and z in D,*

$$\frac{1}{C(m)}|f'(x)| \leq \frac{|f(y) - f(z)|}{|y - z|} \leq C(m)|f'(x)|.$$

Proof. If $U = \mathbb{C}$ then f is an affine map and the theorem is immediate with $C(m) = 1$. Otherwise, let $g : (\Delta, 0) \to (U, x)$ be a Riemann mapping. Then $\mathrm{mod}(g^{-1}(D), \Delta) = \mathrm{mod}(D, U) > m > 0$, so there is an $r(m) < 1$ such that $g^{-1}(D) \subset \Delta(r(m))$ by Theorem 2.4. Now apply the Koebe theorem for univalent maps on the unit disk to g and $f \circ g$.

■

2.5 Normal families

Definition. Let X be a complex manifold, and let $\mathcal{F}$ be a family of holomorphic maps $f : X \to \widehat{\mathbb{C}}$. Then $\mathcal{F}$ is a *normal family* if every sequence f_n in $\mathcal{F}$ has a subsequence which converges uniformly on compact subsets of X. The limit f_∞ is again a holomorphic map to $\widehat{\mathbb{C}}$.

Theorem 2.10 (Montel) *For any complex manifold, the set of all holomorphic maps into $\widehat{\mathbb{C}} - \{0, 1, \infty\}$ is a normal family.*

The proof is based on the Schwarz Lemma and the fact that the triply-punctured sphere is covered by the unit disk.

Montel's theorem is one of the basic tools used in the classical theory of iterated rational maps developed by Fatou and Julia. It is easy to see that any three distinct points on the Riemann sphere can replace the triple $\{0, 1, \infty\}$ in the statement of the theorem. More generally, we have:

Corollary 2.11 *Let $s_i : X \to \widehat{\mathbb{C}}$, $i = 1, 2, 3$ be three holomorphic maps whose graphs are disjoint. Then the set $\mathcal{F}$ of all holomorphic maps $f : X \to \widehat{\mathbb{C}}$ whose graphs are disjoint from the graphs of $\{s_1, s_2, s_3\}$ is a normal family.*

Proof. There is a holomorphically varying Möbius transformation $A(x)$ mapping $\{s_1(x), s_2(x), s_3(x)\}$ to $\{0, 1, \infty\}$. A sequence f_n in $\mathcal{F}$ determines a sequence $g_n(x) = A(x)(f_n(x))$ mapping X into the sphere and omitting the values 0, 1 and ∞. Thus g_n has a convergent subsequence g_{n_k}, so f_n has a convergent subsequence $f_{n_k}(x) = A(x)^{-1}(g_{n_k}(x))$.

∎

See [Bea2, §3.3], [Mon].

2.6 Quasiconformal maps

We will have occasional need for the theory of quasiconformal maps; basic references for the facts summarized below are [AB], [Ah1] and [LV].

Definition. A homeomorphism $f : X \to Y$ between Riemann surfaces X and Y is *K-quasiconformal*, $K \geq 1$ if for all annuli $A \subset X$,

$$\frac{1}{K}\operatorname{mod}(A) \leq \operatorname{mod}(f(A)) \leq K \operatorname{mod}(A).$$

This is equivalent to the following analytic definition: f is K-quasiconformal if locally f has distributional derivatives in L^2, and if the *complex dilatation* μ, given locally by

$$\mu(z)\frac{d\bar{z}}{dz} = \frac{\bar{\partial}_z f}{\partial_z f} = \frac{\partial f/\partial \bar{z}}{\partial f/\partial z}\frac{d\bar{z}}{dz},$$

satisfies $|\mu| \leq (K-1)/(K+1)$ almost everywhere. Note that μ is an L^∞ *Beltrami differential*, that is a form of type $(-1, 1)$.

A mapping f is 1-quasiconformal if and only if f is conformal in the usual sense.

The great flexibility of quasiconformal maps comes from the fact that any μ with $\|\mu\|_\infty < 1$ is realized by a quasiconformal map. This is the "measurable Riemann mapping theorem":

Theorem 2.12 (Ahlfors-Bers) *For any L^∞ Beltrami differential μ on the plane with $\|\mu\|_\infty < 1$, there is a unique quasiconformal map $\phi : \mathbb{C} \to \mathbb{C}$ such that ϕ fixes 0 and 1 and the complex dilatation of ϕ is μ.*

Moreover, for any μ with $\|\mu\|_\infty \leq 1$, we may construct a family of quasiconformal maps $\phi_t : \mathbb{C} \to \mathbb{C}$, $|t| < 1$, satisfying

$$\frac{\overline{\partial}_z \phi_t}{\partial_z \phi_t} = t\mu$$

and normalized as above. Then $\phi_t(z)$ is a holomorphic function of $t \in \Delta$ for each $z \in \mathbb{C}$.

A mapping preserving the measurable complex structure specified by μ can be viewed as holomorphic after a quasiconformal change of coordinates. Here is an application to rational maps that we will use in §4.2:

Theorem 2.13 *Let $f : \widehat{\mathbb{C}} \to \widehat{\mathbb{C}}$ be a rational map, and let μ be a Beltrami differential on the sphere such that $f^*\mu = \mu$ and $\|\mu\|_\infty < 1$. Then $g = \phi \circ f \circ \phi^{-1}$ is also a rational map, where ϕ is any quasiconformal map with complex dilatation μ.*

Proof. Using the chain rule one may check that g is 1-conformal, hence holomorphic, away from its branch points. The latter are removable singularities.

∎

This principle forms the basis for the no wandering domains theorem and for the Teichmüller theory of rational maps [Sul3], [McS].

2.7 Measurable sets

The small scale geometry of a measurable set is controlled by:

Theorem 2.14 (Lebesgue density) *Let $E \subset \widehat{\mathbb{C}}$ be a measurable set of positive area. Then*

$$\lim_{r \to 0} \frac{\text{area}(E \cap B(x,r))}{\text{area } B(x,r)} = 1$$

for almost every x in E.

See, e.g. [Stein, §I.1]. Here $B(x,r)$ is a ball about x of radius r in the spherical metric, and area denotes spherical area. Any two smooth metrics in the same conformal class result in the same limit above.

Corollary 2.15 *Let $f : \widehat{\mathbb{C}} \to \mathbb{R}^n$ be a measurable function. Then for all $\epsilon > 0$ and almost every x in $\widehat{\mathbb{C}}$,*

$$\lim_{r \to 0} \frac{\text{area}(\{y \in B(x,r) \ : \ |f(y) - f(x)| < \epsilon\})}{\text{area } B(x,r)} = 1.$$

When the limit above is equal to one for every $\epsilon > 0$, we say f is *almost continuous* at x.

2.8   Absolute area zero

It is sometimes useful to study a compact set $F \subset \mathbb{C}$ in terms of the Riemann surface $X = \mathbb{C} - F$. In this section we give a criterion for F to be a set of area zero, using the conformal geometry of X.

Definitions. The set F is of *absolute area zero* if the area of $\mathbb{C} - f(X)$ is zero for any injective holomorphic map $f : X \to \mathbb{C}$. In terms of the classification of Riemann surfaces, this is equivalent to the condition that X is in O_{AD} [SN, p.3].

Since our area criterion will depend only on the conformal geometry of X, it will also show F is of absolute area zero.

A set A is *nested* inside an annulus $B \subset \mathbb{C}$ if A lies in the bounded component of $\mathbb{C} - B$.

Theorem 2.16 *Suppose $E_1, E_2, \ldots$ is a sequence of disjoint open sets in the plane, such that*

1. *E_n is a finite union of disjoint unnested annuli of finite moduli;*

2. *any component A of E_{n+1} is nested inside some component B of E_n; and*

3. *for any sequence of nested annuli A_n, where A_n is a component of E_n, we have $\sum \text{mod}(A_n) = \infty$.*

Let F_n be the union of the bounded components of $\mathbb{C} - E_n$, and let $F = \bigcap F_n$. Then F is a totally disconnected set of absolute area zero.

The set F consists of those points which are nested inside infinitely many components of $\bigcup E_n$.

Lemma 2.17 *Let $U \subset \mathbb{C}$ be a disk of finite area, let $K \subset U$ be a connected compact set, and let A be the annulus $U - K$. Then*

$$\text{area}(K) \le \frac{\text{area}(U)}{1 + 4\pi \bmod(A)}.$$

Proof. Let Γ be the collection of simple closed curves in A which represent the generator of $\pi_1(A)$. By the method of extremal length, the modulus of A satisfies

$$\bmod(A) \le \frac{\int_A \rho^2(z)|dz|}{\inf_{\gamma \in \Gamma} \left(\int_\gamma \rho(z)|dz| \right)^2}$$

for any finite area conformal metric $\rho(z)|dz|$ on A [Ah1, p.13]. Taking ρ to be the Euclidean metric, the numerator above becomes $\text{area}(A)$, while the isoperimetric inequality gives $(\int_\gamma |dz|)^2 \ge 4\pi\,\text{area}(K)$ for every γ in Γ. Since $\text{area}(A) = \text{area}(U) - \text{area}(K)$, we have

$$\bmod(A) \le \frac{\text{area}(U) - \text{area}(K)}{4\pi\,\text{area}(K)},$$

and the proof is completed by algebra.

■

Proof of Theorem 2.16. Form a tree (or forest) whose vertices are the components of $\bigcup E_n$ and whose edges join $A \subset E_n$ to $B \subset E_{n+1}$ if B is nested inside A. If we weight each vertex A by $\bmod(A)$, then the sum of the weights along any branch leading to infinity is infinite. Since the tree has finite degree, it follows that $M_n \to \infty$, where

$$M_n = \inf_{\mathcal{A}_n} \sum_1^n \bmod(A_i)$$

and $\mathcal{A}_n$ denotes the collection of all sequences of nested annuli A_1, $\ldots$, A_n such that A_i is a component of E_i.

Using the area-modulus estimate above, one may prove by induction that

$$\text{area}(F_n) \le \text{area}(F_1) \sup_{A_n} \prod_1^n \frac{1}{1 + 4\pi \operatorname{mod}(A_i)},$$

which tends to zero because M_n tends to infinity. Thus $\text{area}(F) = 0$.

If $f : \mathbb{C} - F \to \mathbb{C}$ is a univalent map, then we may apply the same argument to $f(E_i)$ to show the complement of the image of f also has area zero. Therefore F has absolute area zero.

Since any component K of F lies in a descending nest of annuli with $\sum \operatorname{mod}(A_n) = \infty$, K is a point and therefore F is totally disconnected.

■

Remark. We first formulated this criterion for application to cubic polynomials in [BH, §5.4]; compare [Mil3]. Lyubich applies the same criterion to quadratic polynomials in [Lyu4]. A related result appears in [SN, §I.1.D].

2.9   The collar theorem

Let $S(x)$ be the function

$$S(x) = \sinh^{-1}(1/\sinh(x/2)).$$

For a simple geodesic α on a hyperbolic surface, the *collar* about α is given by

$$C(\alpha) \;\; = \;\; \{x \; : \; d(x, \alpha) < S(\ell(\alpha))\}$$

where $d(\cdot)$ denotes the hyperbolic metric.

The following result is due to Buser [Bus1].

Theorem 2.18 (Collars for simple geodesics) *The collar $C(\alpha)$ about a simple geodesic on a hyperbolic surface is an embedded annulus.*

If α and β are disjoint simple geodesics, then $C(\alpha)$ and $C(\beta)$ are disjoint.

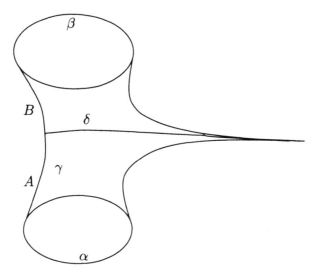

Figure 2.3. Distance between simple geodesics.

Proof. For the first part, pass to the universal cover $\mathbb{H}$ of X, let $\tilde{\alpha}$ be a lift of α, and let $g \in \pi_1(X)$ be a hyperbolic isometry generating the stabilizer of $\tilde{\alpha}$. If the collar $C(\alpha)$ is not embedded, then there is a point $x \in \tilde{\alpha}$ and an $h \in \pi_1(X)$ such that $d(x, hx) < 2S(x)$ and h does not lie in the cyclic group generated by g. By a trigonometry argument,

$$\sinh(d(x, gx)/2) \sinh(d(x, hx)/2) \geq 1,$$

[Beal, Theorem 8.3.1], which is impossible because

$$\sinh(\ell(\alpha)/2) \sinh(S(\ell(\alpha))) = 1.$$

Now let α and β be disjoint simple closed curves; to verify the second part we will show $d(\alpha, \beta) \geq S(\ell(\alpha)) + S(\ell(\beta))$.

Let γ be a geodesic segment of minimal length joining α to β. We may replace X by the covering space Y corresponding to $\pi_1(\alpha \cup \beta \cup \gamma)$, which is a pair of pants. Two ends of Y correspond to α and β; since inclusions are contracting, it suffices to prove the inequality when the third end is a cusp. Let δ be the simple geodesic starting and ending in the cusp. Then δ cuts γ into two segments of length A and B (see Figure 2.3). We can construct a quadrilateral with three right angles and one ideal vertex, whose side lengths are $(\ell(\alpha/2), A, \infty, \infty)$. For such a quadrilateral,

$$\sinh(\ell(\alpha)/2)\sinh(A) = 1$$

[Bea1, Theorem 7.17.1]. Thus $A = S(\ell(\alpha))$. A similar argument gives $B = S(\ell(\beta))$, and $A + B = \ell(\gamma) = d(\alpha, \beta)$.

■

Theorem 2.19 *The modulus of the collar satisfies*

$$\operatorname{mod} C(\alpha) = M(\ell(\alpha)) > 0,$$

where $M(x)$ decreases continuously from infinity to zero as x increases from zero to infinity.

Proof. Since the width of $C(\alpha)$ decreases as the length of α increases, the modulus $M(x)$ is a decreasing function. Its limiting behavior follows from the behavior of $S(x)$.

■

Definition. A *cusp* is a finite volume end of a (noncompact) hyperbolic surface.

A cusp is like a neighborhood of a simple geodesic whose length has shrunk to zero. As the length of a geodesic γ tends to zero, each boundary component of the collar $C(\gamma)$ tends to a horocycle of length 2. A limiting version of the Collar Theorem 2.18 yields:

Theorem 2.20 (Collars for cusps) *Every cusp κ of a hyperbolic surface X has a collar neighborhood $C(\kappa) \subset X$ isometric to the quotient of the region*

$$\{z \ : \ \operatorname{Im}(z) > 1\} \subset \mathbb{H}$$

by the translation $z \mapsto z + 2$.

The collars about different cusps are disjoint, and $C(\kappa)$ is disjoint from the collar $C(\gamma)$ about any simple geodesic γ on X.

Definition. The *injectivity radius* $r(x)$ at a point in a hyperbolic surface X is the radius of the largest embedded hyperbolic ball centered at x. Equivalently, $2r(x)$ is the length of the shortest essential loop on X passing through x.

Theorem 2.21 (Thick-thin decomposition) *Let X be a hyperbolic surface. There is a universal $\epsilon_0 > 0$ such that all simple geodesics of length less than ϵ_0 are disjoint, and every point x with injectivity radius less than $\epsilon_0/2$ belongs to the collar neighborhood of a unique cusp or short geodesic.*

Proof. As the length $\ell(\gamma)$ of a geodesic γ tends to zero the distance between the boundary components of its collar $C(\gamma)$ tends to infinity, so all sufficiently short geodesics are disjoint. Through any point $x \in X$ there is a simple essential loop of length $2r(x)$, isotopic to a unique cusp or geodesic on X. Since the injectivity radius is bounded below near the boundary of the collar about a short geodesic or cusp, x itself belongs to the interior of the corresponding collar when $r(x)$ is sufficiently small.

∎

For more details see [Bus2, Ch.4, §4.4], [BGS], and [Yam].

Corollary 2.22 *There is a universal $C > 0$ such that for any two points x and y on a hyperbolic surface X, the injectivity radius satisfies*

$$|\log r(x) - \log r(y)| \;\leq\; Cd(x,y).$$

In other words, the log of the injectivity radius is uniformly Lipschitz.

Proof. It is obvious that $r(x)$ is Lipschitz with constant 1, so $\log r(x)$ is Lipschitz if $r(x)$ is not too small. But when $r(x)$ is small, x belongs to a standard collar by the thick-thin decomposition, and there the Lipschitz property can be verified directly.

∎

We conclude this section with an estimate of the distance of a smooth loop from its geodesic representative.

Theorem 2.23 *Let X be a hyperbolic surface, and let x be a point on a loop $\delta \subset X$ which is homotopic to a geodesic γ. Then:*

$$\cosh^2(d(x,\gamma)) \;\leq\; \frac{\cosh^2(\ell(\delta)/2) - 1}{\cosh^2(\ell(\gamma)/2) - 1}.$$

In particular, a lower bound on $\ell(\gamma)$ and an upper bound on $\ell(\delta)$ gives an upper bound on the distance from x to γ.

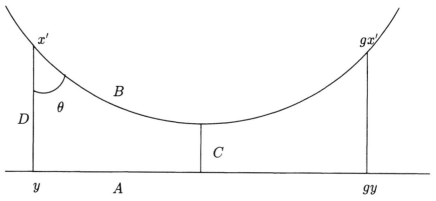

Figure 2.4. Distance to a geodesic.

Proof. Let $X = \mathbb{H}/\pi_1(X)$ present X as a quotient of the hyperbolic plane by a discrete group of isometries. Choose a lift of γ to a geodesic γ' in $\mathbb{H}$, and a compatible lift of x to a point x' (using the homotopy from δ to γ). Then there is a $g \in \pi_1(X)$ stabilizing γ' and translating it distance $\ell(\gamma)$, and $d(x', gx') \leq \ell(\delta)$ because x' and gx' are connected by a lift of δ.

Let y and gy be the points nearest to x' and gx' on $\gamma' \subset \mathbb{H}$. Taking the perpendicular bisector of the geodesic segment from y to gy, we can form a quadrilateral with three right angles, three sides of length $A = \ell(\gamma)/2$, $B = d(x', gx')/2 \leq \ell(\delta)/2$, and $D = d(x', \gamma) = d(x, \gamma)$, and angle θ between sides B and D (see Figure 2.4). By hyperbolic trigonometry [Beal, §7.17], we have the relations

$$\begin{aligned} \sin\theta &= \cosh A / \cosh B \\ \sin\theta &= \cosh C / \cosh D \\ \cos\theta &= \sinh C \sinh B; \end{aligned}$$

squaring and solving for $\cosh^2(D)$ gives the theorem.

∎

2.10 The complex shortest interval argument

Any finite collection of disjoint intervals on the real line contains a shortest member I. In real dynamics one may capitalize on the fact that I is shorter than its neighboring intervals; for example, this fact will be used in §11, and it appears in many other one-dimensional arguments.

In this section we establish a result about hyperbolic surfaces inspired by this shortest interval argument.

Theorem 2.24 (Complex shortest interval) *Let X be a finitely connected planar hyperbolic surface with one cusp, whose remaining ends are cut off by geodesics $\gamma_1, \ldots \gamma_n$, $n > 1$. Suppose the length of every γ_i is greater than $L > 0$. Then there are two distinct geodesics such that*

$$d(\gamma_j, \gamma_k) \leq D(L).$$

Proof. Let X' be the complete surface with geodesic boundary obtained as the closure of the finite volume component of $X - \bigcup \gamma_i$. By the Gauss-Bonnet theorem, the hyperbolic area of X' is $-2\pi\chi(X') = 2\pi(n-1)$. We will construct disjoint neighborhoods E_i of γ_i whose area can be estimated.

Let D be the minimum distance between any two geodesics among $\langle \gamma_i \rangle$. By the thick-thin decomposition (Theorem 2.21) there is an $\epsilon_0 > 0$ such that any loop of length less than ϵ_0 on a hyperbolic surface lies in a collar neighborhood of a unique simple geodesic or cusp. Let $\epsilon = \min(\epsilon_0, L, S^{-1}(D/2))$, where $S(x)$ is the function which appears in the collar lemma (see §2.9). Let Σ be the union of the simple geodesics of length less than ϵ.

The E_i are constructed as follows.

(a) If there is a component of $X' - \Sigma$ containing a unique curve γ_i, set E_i equal to this component.

(b) Otherwise, let $E_i = C_i(D/2)$ where

$$C_i(r) = \{x \in X' : d(x, \gamma_i) < r\}.$$

In case (a), E_i is a complete hyperbolic surface with geodesic boundary, so $\mathrm{area}(E_i) \geq 2\pi$.

In case (b), note that for $0 < r < D/2$, the length of $\partial_0 C_i(r) = \partial C_i(r) - \gamma_i$ is greater than ϵ. For otherwise every component of $\partial_0 C_i(r)$ is homotopic to a short geodesic or the unique cusp of X', and we are in case (a). Since

$$\frac{d\,\mathrm{area}(C_i(r))}{dr} = \mathrm{length}(\partial_0 C_i(r)),$$

we have $\mathrm{area}(E_i) \geq \epsilon D/2$ in case (b).

The regions E_i obtained in this way are disjoint. Indeed, it is clear that two regions of type (a) cannot meet, nor can two regions of type (b), since $d(\gamma_i, \gamma_j) \geq D$. Finally a region of type (b) cannot meet one of type (a), because every curve in Σ is distance at least $D/2$ from every γ_i. This follows from the Collar Theorem 2.18 and the fact $S(\epsilon) \geq D/2$.

Therefore $\sum_1^n \mathrm{area}(E_i) \leq \mathrm{area}(X') = 2\pi(n-1)$. Consequently at least one E_i is of type (b), with $2\pi \geq \mathrm{area}(E_i) \geq D\epsilon/2$, so $D \leq 4\pi/\epsilon$. Since ϵ only depends on L, the theorem follows.

■

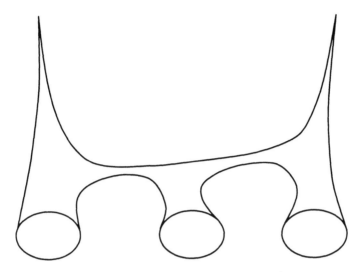

Figure 2.5. All boundary components are far apart.

Remarks.

1. The importance of the preceding result is that the bound on $d(\gamma_i, \gamma_j)$ is independent of n.

2. This result is related to the real shortest interval argument as follows. Suppose $X = \mathbb{C} - \bigcup_1^n I_i$, where I_i are disjoint closed intervals on the real axis. Then the geodesics γ_j and γ_k enclosing the shortest interval and one of its neighbors will be a bounded distance apart whenever we have a lower bound on $\ell(\gamma_i)$.

3. An alternate approach to the result above is to realize X as the complement of a finite set of round disks $D_1, \ldots D_n$ in $\mathbb{C}$ (any finitely connected planar surface with one cusp can be so realized — see [Bie, p.221]). Then γ_j and γ_k can be chosen as the geodesics enclosing D_j and D_k, the disk of smallest diameter and its nearest neighbor.

4. The result fails if X is allowed to have two or more cusps (see Figure 2.5).

2.11   Controlling holomorphic contraction

Definitions. Let $f : X \to Y$ be a holomorphic map between hyperbolic Riemann surfaces. Let $\|f'\|$ denote the norm of the derivative with respect to the hyperbolic metrics on domain and range, and define the *real log derivative* of f by

$$Df(x) = \log \|f'(x)\|.$$

By the Schwarz Lemma, $\|f'(x)\| \leq 1$ so $Df(x) \leq 0$. The function Df is an additive cocycle in the sense that

$$D(f \circ g)(x) = Dg(x) + Df(g(x)).$$

If $f'(x) = 0$ we set $Df(x) = -\infty$.

In this section we bound the variation of $\|f'(x)\|$ (or equivalently $Df(x)$) as the point x varies. To this end it is useful to introduce the 1-form

$$Nf(x) = d(Df(x)),$$

the *real nonlinearity* of f, which measures the infinitesimal variation of $\|f'\|$. Then for any x_1 and x_2 in X, we have

$$|Df(x_1) - Df(x_2)| \;\leq\; \left| \int_\gamma Nf(z)|dz| \right| \;\leq\; d(x_1, x_2) \sup_\gamma \|Nf\|,$$

where γ is a minimal geodesic joining x_1 to x_2 and $\|Nf\|$ denotes the norm of the real nonlinearity measured in the hyperbolic metric on X.

Example. Let $f : \Delta \to \Delta$ be a holomorphic map with $f(0) = 0$. Then an easy calculation shows:

$$Df(0) = \log |f'(0)| \quad \text{and} \quad \|Nf(0)\| = \left| \frac{f''(0)}{2f'(0)} \right|.$$

For our applications the most important case is that of an inclusion $f : X \hookrightarrow Y$. We begin by showing $\|f'(x)\|$ is small if x is close to the boundary of X in Y.

Theorem 2.25 *Let $f : X \subset Y$ be an inclusion of one hyperbolic Riemann surface into another, and let $s = d(x, Y - X)$ in the hyperbolic metric on Y. Then if $s < 1/2$ we have*

$$\|f'(x)\| = O(|s \log s|).$$

In particular $\|f'(x)\| \to 0$ as $s \to 0$.

Proof. By the Schwarz lemma we can reduce to the extremal case $Y = \Delta$, $X = \Delta^*$, $x > 0$ and $s = d(0, x)$ in the hyperbolic metric on Δ. As $s \to 0$ we have $x \sim s/2$ and

$$\|f'(x)\| = \rho_\Delta(x)/\rho_{\Delta^*}(x) = \frac{2|x \log x|}{1 - x^2} \sim |s \log s|,$$

where ρ_Δ and ρ_{Δ^*} are the hyperbolic metrics on the disk and punctured disk.

∎

Now we turn to the variation of $\|f'(x)\|$.

Theorem 2.26 *Let $f : X \to Y$ be a holomorphic map between hyperbolic Riemann surfaces such that f' is nowhere vanishing. Then*

$$\|Nf(x)\| = O(|Df(x)|).$$

Proof. Passing to the universal covers of domain and range, it suffices to treat the case where $X = Y = \Delta$, $x = 0$ and $f : \Delta \to \Delta$ is a holomorphic map without critical points such that $f(0) = 0$. Since $Df(x) = \log|f'(0)|$ and $\|Nf(x)\| = |f''(0)|/(2|f'(0)|)$, we are seeking a bound of the form

$$|f''(0)| \le C|f'(0) \log|f'(0)||.$$

We treat two cases, depending on whether or not $|f'(0)|$ is close to one.

First we write $f(z) = zg(z)$, where $g : \Delta \to \overline{\Delta}$ is also holomorphic, $g(0) = f'(0)$ and $f''(0) = 2g'(0)$. By the Schwarz lemma applied to g, we obtain

$$|f''(0)| \;=\; |2g'(0)| \;\le\; 2(1 - |g(0)|^2) \;=\; 2(1 - |f'(0)|^2).$$

For $1/2 \leq x \leq 1$ we have $1 - x^2 = O(|x \log(x)|)$, so this bound is of the required form when $|f'(0)| \geq 1/2$.

Now we treat the case when $|f'(0)|$ is small; here we will use the fact that f' is nonvanishing.

By the Schwarz lemma applied to f, we have

$$|f'(z)| \leq \frac{1 - |f(z)|^2}{1 - |z|^2} \leq \frac{4}{3}$$

for $z \in \Delta(1/2)$, the disk of radius $1/2$ centered at the origin. Since f' is nonvanishing, it restricts to a map $f' : \Delta(1/2) \rightarrow \Delta(4/3) - \{0\}$. Thus we obtain a holomorphic map $h : \Delta \rightarrow \Delta^*$ by setting $h(z) = (3/4)f'(z/2)$. Since the hyperbolic metric on the punctured disk Δ^* is given by $|dz|/|z \log |z||$, the Schwarz lemma applied to h yields

$$|h'(0)| = \frac{3}{8}|f''(0)| \leq 2|h(0) \log |h(0)|| = \frac{3}{2}\left|f'(0) \log \frac{3}{4}|f'(0)|\right|.$$

For $0 < x < 1/2$ we have $|\log(3x/4)| = O(|\log x|)$, so this bound is of the desired form when $|f'(0)| < 1/2$. Combining these two cases we obtain the theorem.

∎

Integrating this bound, we obtain:

Corollary 2.27 (Variation of contraction) *For any two points* $x_1, x_2 \in X$,
$$\|f'(x_1)\|^{1/\alpha} \geq \|f'(x_2)\| \geq \|f'(x_1)\|^{\alpha}$$
where $\alpha = \exp(Cd(x_1, x_2))$ *for a universal constant* $C > 0$, *and* $d(\cdot)$ *denotes the hyperbolic metric on* X.

Proof. By the preceding theorem, the norm of the one-form
$$\frac{Nf(x)}{Df(x)} = d \log |Df(x)|$$
is bounded by a universal constant with respect to the hyperbolic metric on X. Thus
$$|\log |Df(x_1)| - \log |Df(x_2)|| \leq Cd(x_1, x_2),$$
which is equivalent to the Corollary. ∎

We can summarize these bounds by saying that for any holomorphic immersion $f : X \to Y$ between hyperbolic Riemann surfaces,

$$\log \log \left(\frac{1}{\|f'(x)\|} \right)$$

is a Lipschitz function on X with uniform Lipschitz constant. In particular, if f is only moderately contracting at $x \in X$, then f is not very contracting within a bounded distance of x.

A prototypical example is provided by the inclusion $f : \Delta^* \hookrightarrow \Delta$; as z tends to zero along a hyperbolic geodesic in Δ^*, $\log \log(1/\|f'(z)\|)$ grows approximately linearly with respect to distance along the geodesic, so the bounds above are the right order of magnitude.

Next we will show for an arbitrary inclusion $f : X \hookrightarrow Y$, the bounds above can be improved on the thick part of X. In other words, the rapid variation of f' for the map $\Delta^* \hookrightarrow \Delta$ is accounted for by the small injectivity radius near the cusp.

Theorem 2.28 *Let $f : X \hookrightarrow Y$ be an inclusion between hyperbolic Riemann surfaces. Then*

$$\|Nf(x)\| = O\left(\frac{1}{\min(1, r(x))} \right),$$

where $r(x)$ is the injectivity radius of X at x. In particular, a lower bound on $r(x)$ gives an upper bound on $\|Nf(x)\|$.

Proof. As before we pass to universal covers of domain and range and normalize so $x = f(x) = 0$; then we obtain a map $f : \Delta \to \Delta$ such that f is injective on the hyperbolic ball B of radius $r(x)$ about the origin. We have $B = \Delta(s)$ where $s \asymp r(x)$ when $r < 1$. The map $h : \Delta \to \Delta$ given by $h(z) = f(sz)$ is univalent. By Koebe compactness of univalent maps, $|h''(0)/h'(0)| < C$ for a universal constant C. Since $h''(0)/h'(0) = sf''(0)/f'(0)$, we obtain $\|Nf(x)\| \leq C/s = O(1/r(x))$ when $r(x) < 1$.

When $r(x) > 1$ the same argument gives $\|Nf(x)\| = O(1)$.

∎

For our applications the qualitative version below is easiest to apply. Note that this Corollary improves Corollary 2.27 when $\|f'(x_1)\|$ is small.

Corollary 2.29 *Let $f : X \hookrightarrow Y$ be an inclusion between hyperbolic surfaces. Then for any x_1 and x_2 in X,*

$$\frac{1}{C(r,d)} \leq \frac{\|f'(x_1)\|}{\|f'(x_2)\|} \leq C(r,d)$$

where $C(r,d) > 0$ is a continuous function depending only on the injectivity radius $r = r(x_1)$ and the distance $d = d(x_1, x_2)$ between x_1 and x_2.

Proof. Let γ be a path of length $d(x_1, x_2)$ joining x_1 to x_2. By Corollary 2.22, the injectivity radius $r(x)$ is bounded below along γ in terms of $d(x_1, x_2)$ and $r(x_1)$. By the preceding result, we obtain an upper bound on $\|Nf(x)\|$ along γ. The integral of this bound controls $|Df(x_1) - Df(x_2)|$, and thereby the ratio $\|f'(x_1)\|/\|f'(x_2)\|$.

■

Chapter 3

Dynamics of rational maps

This chapter reviews well-known features of the topological dynamics of rational maps, and develops general principles to study their measurable dynamics as well.

We first recall some basic results in rational dynamics (§3.1). A rational map f of degree greater than one determines a partition of the Riemann sphere into a pair of totally invariant sets, the Julia set $J(f)$ and the Fatou set $\Omega(f)$. The behavior of f on the Fatou set is well understood: every component eventually cycles, and the cyclic components are the basins of attracting or parabolic cycles, or rotation domains (Siegel disks or Herman rings).

The dynamics on the Julia set is more mysterious in general. For example, we do not know if f is ergodic whenever the Julia set is equal to the whole Riemann sphere. We will see, however, that an important role is played by the *postcritical set* $P(f)$, defined as the closure of the forward orbits of the critical points.

In §3.2, we use the hyperbolic metric on $\widehat{\mathbb{C}} - P(f)$ to establish expanding properties of f outside of the postcritical set. In §3.3 this expansion leads to the following dichotomy: a rational map either acts ergodically on the sphere, or its postcritical set behaves as a measure-theoretic attractor. The main idea of §3.3 appears in [Lyu1].

Hyperbolic rational maps are introduced in §3.4, and we use the results just developed to show their Julia sets have measure zero.

In §3.5 we turn to an analysis of invariant measurable line fields supported on the Julia set. We first present the known examples of rational maps admitting invariant line fields, namely those which are covered by integral torus endomorphisms. (Examples of this type are due to Lattès [Lat].) Then we show for any other example, the postcritical set must act a measure-theoretic attractor for points in the support of the line field.

This conclusion will later form the first step in our proof that a robust quadratic polynomial is rigid.

3.1   The Julia and Fatou sets

Let $f : \widehat{\mathbb{C}} \to \widehat{\mathbb{C}}$ be a rational map of degree greater than one.

Definitions.   A point z such that $f^p(z) = z$ for some $p \geq 1$ is a *periodic point* for f. The least such p is the *period* of z. If $f^i(z) = f^j(z)$ for some $i > j > 0$ we say z is *preperiodic*.

A *periodic cycle* $A \subset \widehat{\mathbb{C}}$ is a finite set such that $f|A$ is a transitive permutation. The forward orbit of a periodic point is a periodic cycle.

The *multiplier* of a point z of period p is the derivative $(f^p)'(z)$ of the first return map. The multiplier provides a first approximation to the local dynamics of f^p. Accordingly, we say z is

$$
\begin{array}{ll}
repelling & \text{if } |(f^p)'(z)| > 1; \\
indifferent & \text{if } |(f^p)'(z)| = 1; \\
attracting & \text{if } |(f^p)'(z)| < 1; \text{ and} \\
superattracting & \text{if } (f^p)'(z) = 0.
\end{array}
$$

An indifferent point is *parabolic* if $(f^p)'(z)$ is a root of unity.

Remark.   By the definition above, attracting includes superattracting as a special case. This convention is not uniformly adopted in the literature on rational maps, but it is convenient for our purposes.

The *Fatou set* $\Omega(f) \subset \widehat{\mathbb{C}}$ is the largest open set such that the iterates $\{f^n|\Omega : n \geq 1\}$ form a normal family.

The *Julia set* $J(f)$ is the complement of the Fatou set.

The Julia and Fatou sets are each *totally invariant* under f; that is, $f^{-1}(J(f)) = J(f)$ and $f^{-1}(\Omega(f)) = \Omega(f)$; so the partition $\widehat{\mathbb{C}} = J(f) \sqcup \Omega(f)$ is preserved by the dynamics.

The Julia set is the locus of expanding and chaotic behavior; for example:

Theorem 3.1 *The Julia set is equal to the closure of the set of repelling periodic points. It is also characterized as the minimal closed subset of the sphere satisfying $|J| > 2$ and $f^{-1}(J) = J$.*

On the other hand, a normal family is precompact, so one might imagine that the forward orbit of a point in the Fatou set behaves predictably. Note that f maps each component of the Fatou set to another component. The possible behaviors are summarized in the following fundamental result.

Theorem 3.2 (Classification of Fatou components) *Every component U of the Fatou set is preperiodic; that is, $f^i(U) = f^j(U)$ for some $i > j > 0$. The number of periodic components is finite.*
    A periodic component U, with $f^p(U) = U$, is of exactly one the following types:

1. *An* attracting basin: *there is an attracting periodic point w in U, and $f^{np}(z) \to w$ for all z in U as $n \to \infty$.*

2. *A* parabolic basin: *there is a parabolic periodic point $w \in \partial U$ and $f^{np}(z) \to w$ for all z in U.*

3. *A* Siegel disk: *the component U is a disk on which f^p acts by an irrational rotation.*

4. *A* Herman ring: *the component U is an annulus, and again f^p acts as an irrational rotation.*

Remarks. The classification of periodic components of the Fatou set is contained in the work of Fatou and Julia. The *existence* of rotation domains was only established later by work of Siegel and Herman, while the proof that every component of the Fatou set is preperiodic was obtained by Sullivan [Sul3].
    For details and proofs of the results above, see [McS], [CG] or [Bea2].

Polynomials. Let $f : \widehat{\mathbb{C}} \to \widehat{\mathbb{C}}$ be a *polynomial* map of degree $d > 1$. Then infinity is a superattracting fixed point for f, so the Julia set is a compact subset of the complex plane.

Definition. The *filled Julia set* $K(f)$ is the complement of the basin of attraction of infinity. That is, $K(f)$ consists of those $z \in \mathbb{C}$ such that the forward orbit $f^n(z)$ is bounded.

The Julia set $J(f)$ is equal to the boundary of $K(f)$. By the maximum principle, $\mathbb{C} - K(f)$ is connected.

By the Riemann mapping theorem, one may also establish:

Theorem 3.3 *Let $f(z)$ be a polynomial of degree $d > 1$ with connected filled Julia set $K(f)$. Then there is a conformal map*

$$\phi : (\widehat{\mathbb{C}} - \overline{\Delta}) \to (\widehat{\mathbb{C}} - K(f))$$

such that $\phi(z^d) = f(\phi(z))$. Any other such map is given by $\phi(\omega z)$ where $\omega^{d-1} = 1$.

In particular, ϕ is unique when $d = 2$.

3.2 Expansion

The *postcritical set* $P(f)$ is the closure of the strict forward orbits of the critical points $C(f)$:

$$P(f) = \overline{\bigcup_{c \in C(f), \ n > 0} f^n(c)}.$$

Note that $f(P) \subset P$ and $P(f^n) = P(f)$. The postcritical set is also the smallest closed set containing the critical *values* of f^n for every $n > 0$.

A rational map is *critically finite* if $P(f)$ is a finite set.

A fundamental idea, used repeatedly in the sequel, is that f expands the hyperbolic metric on $\widehat{\mathbb{C}} - P(f)$. This idea is not very useful if $P(f)$ is too big: for example, there exist rational maps with $P(f) = \widehat{\mathbb{C}}$ (see [Rees1], [Rees2]), and even among the quadratic polynomials $f_c(z) = z^2 + c$ we have $P(f_c) = J(f_c)$ for a dense G_δ of c's in the boundary of the Mandelbrot set.

On the other hand, there are interesting circumstances when the postcritical set is controlled; for example, $P(f_c)$ is confined to the real axis when c is real, and we will see that $P(f_c)$ is a Cantor set of measure zero when f_c is robust (§9).

For the hyperbolic metric on $\widehat{\mathbb{C}} - P(f)$ to be defined, it is necessary that the postcritical set contain at least three points. The exceptional cases are handled by the following observation:

Theorem 3.4 *If f is a rational map of degree greater than one and $|P(f)| < 3$, then f is conjugate to z^n for some n and its Julia set is a round circle.*

In particular the Julia set has area zero when $|P(f)| < 3$.

Theorem 3.5 *Let f be a rational map with $|P(f)| \geq 3$. If $x \in \widehat{\mathbb{C}}$ and $f(x)$ does not lie in the postcritical set of f, then*

$$\|f'(x)\| \geq 1$$

with respect to the hyperbolic metric on $\widehat{\mathbb{C}} - P(f)$.

Proof. Let $Q(f) = f^{-1}(P(f))$. Then

$$f : (\widehat{\mathbb{C}} - Q(f)) \rightarrow (\widehat{\mathbb{C}} - P(f))$$

is a proper local homeomorphism, hence a covering map, and therefore f is an isometry between the hyperbolic metrics on domain and range. On the other hand, $P(f) \subset Q(f)$ so there is an inclusion $\iota : (\widehat{\mathbb{C}} - Q(f)) \rightarrow (\widehat{\mathbb{C}} - P(f))$. By the Schwarz lemma, inclusions are contracting, so f is expanding.

∎

It can happen that $\|f'(x)\| = 1$ at some points, for example when f has a Siegel disk.

Theorem 3.6 (Julia expansion) *For every point x in $J(f)$ whose forward orbit does not land in the postcritical set $P(f)$,*

$$\|(f^n)'(x)\| \rightarrow \infty$$

with respect to the hyperbolic metric on $\widehat{\mathbb{C}} - P(f)$.

Proof. Let $Q_n = f^{-n}(P(f))$ be the increasing sequence of compact sets obtained as preimages of $P(f)$. The map

$$f^n : (\widehat{\mathbb{C}} - Q_n) \to (\widehat{\mathbb{C}} - P(f))$$

is a proper local homeomorphism, hence a covering map, so f^n is a local isometry from the Poincaré metric on $\widehat{\mathbb{C}} - Q_n$ to the Poincaré metric on $\widehat{\mathbb{C}} - P(f)$. Since we are assuming $|P(f)| > 2$, Theorem 3.1 implies the Julia set is contained in the closure of the union of the Q_n. Thus the spherical distance $d(Q_n, x) \to 0$ as $n \to \infty$. Then the distance r_n from x to Q_n in the Poincaré metric on $\widehat{\mathbb{C}} - P(f)$ tends to zero as well. By Theorem 2.25, the inclusion

$$\iota_n : (\widehat{\mathbb{C}} - Q_n) \to (\widehat{\mathbb{C}} - P(f))$$

satisfies $\|\iota_n'(x)\| \le C|r_n \log r_n| \to 0$, where the norm of the derivative of ι_n is measured using the Poincaré metrics on its domain and range. It follows that $f^n \circ \iota_n^{-1}$ expands the Poincaré on $\widehat{\mathbb{C}} - P(f)$ at x by a factor greater than $1/(C|r_n \log r_n|) \to \infty$ as $n \to \infty$.

∎

The postcritical set is closely tied to the attracting and indifferent dynamics of f, as demonstrated by the following Corollary (which goes back to Fatou; compare [CG, p.82]).

Corollary 3.7 *The postcritical set $P(f)$ contains the attracting cycles of f, the indifferent cycles which lie in the Julia set, and the boundary of every Siegel disk and Herman ring.*

Proof. The Corollary is immediate for $f(z) = z^n$, so we may assume $|P(f)| > 2$.

Let x be a fixed point of f^p. If x is attracting then $x \in P(f)$ by Theorem 3.5. If x is indifferent and $x \in J(f)$, then $x \in P(f)$ by the preceding result. (Note this case includes all parabolic cycles).

Let K be a component of the boundary of a Siegel disk or Herman ring U of period p. The postcritical set meets U in a finite collection of f^p-invariant smooth circles (possibly including the center of the Siegel disk as a degenerate case). There is a unique component U_0 of $U - P(f)$ such that $K \subset \overline{U_0}$. Let V_0 be the component of $\widehat{\mathbb{C}} - P(f)$

containing U_0. Since $f^p|U_0$ is a rotation, the hyperbolic metric on V_0 is not expanded and thus $f^p(V_0) = V_0$ and V_0 is contained in the Fatou set. Therefore $U_0 = V_0$ and $K \subset \partial V_0 \subset P(f)$.

∎

The results of §2.11 allow one to control the variation of $\|f'\|$ as well. Here is a result in that direction which we will use in §10.

Theorem 3.8 (Variation of expansion) *Let* $f : \widehat{\mathbb{C}} \to \widehat{\mathbb{C}}$ *be a rational map with* $|P(f)| \geq 3$. *Let* γ *be a path joining two points* $x_1, x_2 \in \widehat{\mathbb{C}}$, *such that* $f(\gamma)$ *is disjoint from the postcritical set, and let* d *be the parameterized length of* $f(\gamma)$ *in the hyperbolic metric on* $\widehat{\mathbb{C}} - P(f)$. *Then:*

$$\|f'(x_1)\|^\alpha \geq \|f'(x_2)\| \geq \|f'(x_1)\|^{1/\alpha},$$

where $\alpha = \exp(Cd)$ *for a universal* $C > 0$; *and*

$$\frac{1}{C(r,d)} \leq \frac{\|f'(x_1)\|}{\|f'(x_2)\|} \leq C(r,d),$$

where r *denotes the injectivity radius of* $\widehat{\mathbb{C}} - P(f)$ *at* $f(x_1)$.

Proof. Let $Q(f) = f^{-1}(P(f))$; then

$$f : (\widehat{\mathbb{C}} - Q(f)) \to (\widehat{\mathbb{C}} - P(f))$$

is a covering map, hence a local isometry for the respective hyperbolic metrics, while the inclusion

$$\iota : (\widehat{\mathbb{C}} - Q(f)) \hookrightarrow (\widehat{\mathbb{C}} - P(f))$$

is a contraction. Thus whenever $f(x) \notin P(f)$ we have

$$\|f'(x)\| = \frac{1}{\|\iota'(x)\|},$$

where the latter norm is measured from the hyperbolic metric on the complement of $Q(f)$ to that on the complement of $P(f)$.

Since f is a local isometry, the length of γ in the hyperbolic metric on $\widehat{\mathbb{C}} - Q(f)$ is equal to d; in particular, d bounds the distance between x_1 and x_2. By Corollary 2.27,

$$\|f'(x_2)\| = \frac{1}{\|\iota'(x_2)\|} \leq \frac{1}{\|\iota'(x_1)\|^\alpha} = \|f'(x_1)\|^\alpha,$$

where $\alpha = \exp(Cd)$ for a universal constant C. Interchanging the roles of x_1 and x_2, we obtain the first bound. The second bound follows similarly from Corollary 2.29.

∎

It is also natural to think of this result as controlling $\|(f^{-1})'(y)\|$ as y varies on $\widehat{\mathbb{C}} - P(f)$; the control is then in terms of the distance y moves and the injectivity radius at y.

3.3   Ergodicity

Definition. A rational map is *ergodic* if any measurable set A satisfying $f^{-1}(A) = A$ has zero or full measure in the sphere. In this section we prove:

Theorem 3.9 (Ergodic or attracting) *If f is a rational map of degree greater than one, then*

- *the Julia set is equal to the whole Riemann sphere and the action of f on $\widehat{\mathbb{C}}$ is ergodic, or*

- *the spherical distance $d(f^n x, P(f)) \to 0$ for almost every x in $J(f)$ as $n \to \infty$.*

As a sample application, we have:

Corollary 3.10 *If f is critically finite, then either $J(f) = \widehat{\mathbb{C}}$ and f is ergodic, or f has a superattracting cycle and $J(f)$ has measure zero.*

Proof. Since the postcritical set is finite, every periodic cycle of f is either repelling or superattracting (see Theorem A.6). In particular, the periodic cycles in $P(f) \cap J(f)$ are repelling, so

$$\limsup d(f^n x, P(f)) > 0$$

for all $x \in J(f)$ outside the grand orbit of $P(f)$ (a countable set). Thus the postcritical set does *not* attract a set of positive measure in the Julia set.

If f has no superattracting cycle, then $J(f) = \hat{\mathbb{C}}$ (Theorem A.6), so the first alternative of the theorem above must hold. Otherwise $J(f) \neq \hat{\mathbb{C}}$, so the second alternative must hold vacuously, by $J(f)$ having measure zero.

∎

Remark. It appears to be difficult to construct a Julia set of positive measure which is not equal to the whole sphere; see however [NvS].

Lemma 3.11 *Let $V \subset \hat{\mathbb{C}} - P(f)$ be a connected open set, and let U be a component of $f^{-n}(V)$. Then $f^n : U \to V$ is a covering map.*

In particular, if V is simply-connected, there is a univalent branch of f^{-n} mapping V to U.

Proof. The critical values of f^n lie in $P(f)$, so $f^n : U \to V$ is a proper local homeomorphism, hence a covering map.

∎

Lemma 3.12 *Let $U \subset J(f)$ be a nonempty open subset of the Julia set. Then there is an $n > 0$ such that $f^n(U) = J(f)$.*

See [Mil2, Cor. 11.2], or [EL, Theorem 2.4].

Proof of Theorem 3.9(Ergodic or attracting). We may assume $|P(f)| \geq 3$, for otherwise the Julia set is a circle and its area is zero.

Suppose there is a set E of positive measure in the Julia set for which

$$\limsup d(f^n x, P(f)) > \epsilon > 0.$$

Consider any f-invariant set $F \subset J(f)$ such that $E \cap F$ has positive measure. We will show that $F = \widehat{\mathbb{C}}$, so f is ergodic.

Let $K = \{z : d(z, P(f)) > \epsilon\}$, and let x be a point of Lebesgue density of $E \cap F$. By assumption, there are n_k tending to infinity such that $y_k = f^{n_k}(x) \in K$.

Consider the spherical balls of definite size $B_k = B(y_k, \epsilon/2)$. By Lemma 3.11 above, there is a univalent branch g_k of f^{-n_k} defined on B_k and mapping y_k back to x. Moreover g_k can be extended to a univalent function on the larger ball $B(y_k, \epsilon)$, so by the Koebe principle g_k has bounded nonlinearity on B_k. In particular the area of $C_k = g_k(B_k)$ is comparable to the square of its diameter.

By Theorem 3.6, $\|(f^{n_k})'x\| \to \infty$ with respect to the Poincaré metric on $\widehat{\mathbb{C}} - P(f)$. Since K is compact, the same is true with respect to the spherical metric. Therefore the spherical diameter of C_k tends to zero. Since x is a point of density,

$$\frac{\operatorname{area}(F \cap C_k)}{\operatorname{area}(C_k)} \to 1.$$

But F is f-invariant, so by Koebe distortion the density

$$\frac{\operatorname{area}(F \cap B_k)}{\operatorname{area}(B_k)}$$

of F in B_k tends to one as well.

By compactness of the sphere we may pass to a subsequence such that the balls B_k converge to a limiting ball B in which the density of F is equal to one. Therefore $B \subset F$ (a.e.) and by Lemma 3.12 above, $f^n(B) = \widehat{\mathbb{C}}$ for some $n > 0$. Since F is f-invariant, we find $F = J(f) = \widehat{\mathbb{C}}$ a.e. and therefore f is ergodic.

∎

3.4 Hyperbolicity

In this section we give several equivalent definitions of hyperbolic rational maps, displaying some of the properties that make these dynamical systems especially well-behaved. Then we apply Theorem 3.9(Ergodic or attracting) to show the Julia set of a hyperbolic map has measure zero.

Theorem 3.13 (Characterizations of hyperbolicity) *Let f be a rational map of degree greater than one. Then the following conditions are equivalent.*

1. *The postcritical set $P(f)$ is disjoint from the Julia set $J(f)$.*

2. *There are no critical points or parabolic cycles in the Julia set.*

3. *Every critical point of f tends to an attracting cycle under forward iteration.*

4. *There is a smooth conformal metric ρ defined on a neighborhood of the Julia set such that $\|f'(z)\|_\rho > C > 1$ for all $z \in J(f)$.*

5. *There is an integer $n > 0$ such that f^n strictly expands the spherical metric on the Julia set.*

Definition. The map f is *hyperbolic* if any of the equivalent conditions above are satisfied. A hyperbolic rational map is also sometimes said to be *expanding*, or to satisfy Smale's *Axiom A*.

Proof of Theorem 3.13 (Characterizations of hyperbolicity). If $|P(f)| = 2$ then f is conjugate to z^n and it is trivial to verify that all conditions above are satisfied. So suppose $|P(f)| > 2$.

If $P(f) \cap J(f) = \emptyset$, then there are no critical points or parabolic points in the Julia set (since every parabolic point attracts a.critical point.) By Theorem 3.2 and Corollary 3.7, if there are no critical points or parabolic points in the Julia set, then there are no parabolic basins, Siegel disks or Herman rings, and consequently under iteration every critical point tends to an attracting cycle. Clearly this last condition implies $P(f) \cap J(f) = \emptyset$. Thus $1 \implies 2 \implies 3 \implies 1$.

Assuming case 3, we certainly have $P(f) \cap J(f) = \emptyset$, and moreover $P(f)$ and $Q(f) = f^{-1}(P(f))$ are countable sets with only finitely many limit points. Thus $\hat{\mathbb{C}} - P(f)$ and $\hat{\mathbb{C}} - Q(f)$ are connected, and

$$f : (\hat{\mathbb{C}} - Q(f)) \to (\hat{\mathbb{C}} - P(f))$$

is a covering map, hence an isometry for the respective hyperbolic metrics. Since $|P(f)| > 2$, $Q(f) - P(f)$ is nonempty and so the inclusion

$$\iota : (\hat{\mathbb{C}} - Q(f)) \hookrightarrow (\hat{\mathbb{C}} - P(f))$$

is a contraction ($\|\iota'(z)\| < 1$ for all z in $\widehat{\mathbb{C}} - Q(f)$). Thus f expands the hyperbolic metric on $\widehat{\mathbb{C}} - P(f)$, and the expansion is strict on the Julia set because $J(f)$ is a compact subset of $\widehat{\mathbb{C}} - P(f)$. Thus $3 \implies 4$.

Any two conformal metrics defined near the Julia set are quasi-isometric, and the expansion factor of f^n overcomes the quasi-isometry constant when n is large enough. Thus $4 \implies 5$.

Finally, if f^n expands a conformal metric on the Julia set, then $J(f)$ contains no critical points or parabolic cycles; thus $5 \implies 2$ and we have shown $1 - 5$ are equivalent.

$\blacksquare$

Theorem 3.14 *The Julia set of a hyperbolic rational map has measure zero.*

Proof. Since the Julia set of a hyperbolic rational map contains no critical points, it is not equal to the Riemann sphere. If $J(f)$ were to have positive measure, then by Theorem 3.9, almost every point in $J(f)$ would be attracted to the postcritical set. But then $P(f)$ would meet $J(f)$, contrary to the assumption of hyperbolicity.

$\blacksquare$

Remark. In fact, the Hausdorff dimension δ of the Julia set of a hyperbolic rational map satisfies $0 < \delta < 2$ and the δ-dimensional measure of $J(f)$ is finite and positive; see [Sul2].

From Theorem 3.2 one may immediately deduce:

Corollary 3.15 *The attractor A of a hyperbolic rational map consists of a finite set of cycles which can be located by iterating the critical points of f.*

More precisely, if A denotes the set of limit points of the forward orbits of the critical points of f, then A is a set equal to the set of attracting periodic points of f, and $d(f^n(z), A) \to 0$ for almost every z in $\widehat{\mathbb{C}}$.

3.5   Invariant line fields and complex tori

The measurable dynamics of a rational map can be extended by considering the action of f on various bundles over the sphere. For the theory of quasiconformal rigidity, the action of f on the space of unoriented tangent lines plays an essential role. For example, we will later see that hyperbolic dynamics is dense in the quadratic family if and only if there is no quadratic polynomial with an invariant line field on its Julia set (Corollary 4.10).

All *known* examples of rational maps supporting invariant line fields on their Julia sets come from a simple construction using complex tori. In this section we will show $P(f)$ must attract the support of the line field in any other type of example. This theorem represents an initial step towards proving such additional examples do not exist.

Definition. A *line field* supported on a subset E of a Riemann surface X is the choice of a real line through the origin in the tangent space $T_e X$ at each point of E.

A line field is the same as a Beltrami differential $\mu = \mu(z)d\bar{z}/dz$ supported on E with $|\mu| = 1$. A Beltrami differential determines a function on the tangent space, homogeneous of degree zero, by

$$\mu(v) = \mu(z)\frac{\bar{a}(z)}{a(z)},$$

where $v = a(z)\partial/\partial z$ is a tangent vector. The corresponding line field consists of those tangent vectors for which $\mu(v) = 1$ (union the zero vector). Conversely, the real line through $a\partial/\partial z$ corresponds to the Beltrami differential $(a/\bar{a})d\bar{z}/dz$.

A line field is *holomorphic* (meromorphic) if locally

$$\mu = \bar{\phi}/|\phi|,$$

where $\phi = \phi(z)dz^2$ is a holomorphic (meromorphic) quadratic differential. In this case we say μ is *dual to* ϕ. Note that ϕ is unique up to a positive real multiple.

A line field is *measurable* if $\mu(z)$ is a measurable function.

Let f be a rational map. We say f admits an *invariant line field* if there is a *measurable* Beltrami differential μ on the sphere such that

$f^*\mu = \mu$ a.e., $|\mu| = 1$ on a set of positive measure and μ vanishes elsewhere. We are mostly interested in line fields which are *carried on the Julia set*, meaning $\mu = 0$ outside $J(f)$.

Examples.

1. The radial line field in the plane is invariant under $f(z) = z^n$. This line field is dual to the quadratic differential dz^2/z^2, so it is holomorphic outside of zero and infinity.

2. Let $X = \mathbb{C}/\Lambda$ be a complex torus, and let α be a complex number with $|\alpha| > 1$ such that $\alpha\Lambda \subset \Lambda$. Then multiplication by α induces an endomorphism $F : X \to X$.

Let $\wp : X \to \widehat{\mathbb{C}}$ be an even function ($\wp(-z) = \wp(z)$) presenting X as a twofold branched covering of the Riemann sphere; an example of such a $\wp$ is the Weierstrass function. Since $\alpha(-z) = -\alpha z$, there is an induced rational map f of degree $|\alpha|^2$ on the sphere such that the diagram

$$\mathbb{C}/\Lambda \xrightarrow{z \mapsto \alpha z} \mathbb{C}/\Lambda$$
$$\wp \downarrow \qquad \wp \downarrow$$
$$\widehat{\mathbb{C}} \xrightarrow{\ f\ } \widehat{\mathbb{C}}$$

commutes. (Compare [Lat].)

In this case we say f is *double covered by an endomorphism of a torus*. Since F has a dense set of repelling periodic points, the Julia set of f is the whole sphere.

Now suppose $\alpha = n > 1$ is an integer. Then the postcritical set $P(f)$ coincides with the set of critical values of $\wp$. Since the critical points of $\wp$ are the points of order two on the torus X, $|P(f)| = 4$.

Multiplication by n preserves any family of parallel lines in $\mathbb{C}$, so F admits an invariant line field on X. This line field descends to an f-invariant line field on $\widehat{\mathbb{C}}$ dual to a meromorphic quadratic differential ϕ with simple poles on the postcritical set $P(f)$ and no zeros. Explicitly,

$$\phi = \frac{dz^2}{(z-p_1)(z-p_2)(z-p_3)(z-p_4)}$$

where $P(f) = \{p_1, p_2, p_3, p_4\}$.

A rational map arising in this way is said to be covered by an *integral* torus endomorphism.

In the introduction we formulated the following:

Conjecture 1.4 (No invariant line fields) *A rational map f carries no invariant line field on its Julia set, except when f is double covered by an integral torus endomorphism.*

We will adapt the arguments of the preceding section to give a result supporting this conjecture.

Lemma 3.16 *Let μ be an f-invariant line field which is holomorphic on a nonempty open set contained in the Julia set. Then f is double covered by an integral torus endomorphism.*

Proof. Note that the hypotheses imply the Julia set of f is the whole sphere.

Let μ be dual to a holomorphic quadratic differential ϕ on an open set $U \subset J(f)$, and let z be a point in the Riemann sphere. Then $f^n(u) = z$ for some u in U and $n > 0$ (by Lemma 3.12).

If $(f^n)'(u) \neq 0$, then there is a univalent map $g : V \to U$ defined on a neighborhood V of z such that $f^n \circ g = \mathrm{id}$. Then μ is dual to $g^*\phi$ on V by f-invariance.

If $(f^n)'(u) = 0$, one can similarly define a meromorphic differential ψ near z to which μ is dual. To construct ψ, choose a neighborhood V of z such that a component V' of $f^{-n}(V)$ is contained in U, and let ψ be the pushforward $(f^n)_*\phi$ of ϕ from V' to V. Since μ is dual to ϕ on each sheet of V', it is dual to ψ on V.

Therefore μ is meromorphic on the sphere. Since the sphere is simply-connected, there is a globally defined meromorphic differential ϕ dual to μ. Invariance of μ implies that $f^*\phi = (\deg f)\phi$, since μ determines ϕ up to a positive real multiple.

We claim that ϕ has simple poles and no zeros. Indeed, if $\phi(z) = 0$ then ϕ also vanishes at all preimages of z under f, which is impossible because the zeros of ϕ are discrete. Similarly, if ϕ were to have a pole of order two or more at z, then it would have poles at all preimages of z.

For any meromorphic quadratic differential on the sphere, the number of poles exceeds the number of zeros by four. Therefore ϕ has four simple poles and no zeros.

It is easy to see that the poles of ϕ coincide with the postcritical set $P(f)$. Indeed, if ϕ fails to have a pole at a critical value of f, then it has a zero at the corresponding critical point, which has already been ruled out. Propagating this pole forward, we have poles at all points of $P(f)$. There can be no other poles, because a pole at $z \notin P(f)$ entails poles along the entire backward orbit of z.

The proof is completed using the orbifold associated to f (see §A.3, Theorem A.5). One can check that f is a critically finite map whose orbifold $\mathcal{O}_f$ has signature $(2, 2, 2, 2)$. Then f lifts to an endomorphism $z \mapsto \alpha z$ of the complex torus $X = \mathbb{C}/\Lambda$ obtained as a twofold cover of $\widehat{\mathbb{C}}$ branched over $P(f)$. The endomorphism is integral because μ lifts to an invariant holomorphic line field on X, which in turn lifts to a family of parallel lines on the universal cover $\mathbb{C}$ of X, invariant under multiplication by α.

∎

The main result of this section is:

Theorem 3.17 (Toral or attracting) *Let f be a rational map with an invariant line field on its Julia set $J(f)$. Either*

1. f is double covered by an integral torus endomorphism, or

2. $d(f^n x, P(f)) \to 0$ for almost every x in $J(f)$.

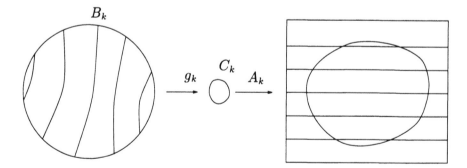

Figure 3.1.   Blowups of an invariant line field.

Proof. Assume there is a set $E \subset J(f)$ of positive measure for which

$$\limsup d(f^n x, P(f)) > \epsilon > 0.$$

Then $J(f) = \hat{\mathbb{C}}$ and f is ergodic, so its invariant line field μ is supported on the full Julia set. We can find a point x in E such that $|\mu(x)| = 1$ and μ is almost continuous at x (Corollary 2.15). This means the line field is nearly parallel on small balls centered at x.

Following the proof of Theorem 3.9, we will use the dynamics to expand the nearly parallel line field up to definite size and thereby make it holomorphic.

As before we first take branches of f^{-n_k} to obtain univalent maps

$$g_k : B_k = B(y_k, \epsilon/2) \to C_k$$

such that $g_k(y_k) = x$, $\mathrm{diam}(C_k) \to 0$, the area of C_k is comparable to its diameter squared and $g_k^*(\mu) = \mu$.

For convenience, choose coordinates on the sphere so that $x = 0$ and $C_k \subset \mathbb{C}$. Next construct linear dilations $A_k(z) = \alpha_k z$, $\alpha_k \to \infty$, such that the composition

$$h_k = A_k \circ g_k : B_k \to \mathbb{C}$$

satisfies $h_k(y_k) = 0$ and $\|h_k'(y_k)\| = 1$ (where the norm is measured from the spherical to the Euclidean metric). Adjusting A_k by a rotation we can also assume that $\mu(x) = (A_k^* \nu)(x)$, where $\nu = d\bar{z}/dz$ is the horizontal line field on the plane (see Figure 3.1).

Since x is a point of almost continuity, μ is nearly equal to $\nu_k = A_k^*(\nu)$ throughout C_k. More precisely, for any $\delta > 0$ the density in C_k of the set where the angle between ν_k and μ exceeds δ tends to zero as $k \to \infty$.

We now appeal to compactness to show that μ is holomorphic on some ball B_∞ of radius $\epsilon/2$. First, pass to a subsequence so that B_k tends to a limiting ball B_∞. The maps h_k are univalent functions, so by the Koebe principle there is a further subsequence for which h_k converges uniformly on compact sets to a univalent map $h_\infty : B_\infty \to \mathbb{C}$.

We claim that $\mu = h_\infty^*(\nu)$ on B_∞. Indeed, for large k, μ and $A_k^*(\nu)$ are nearly aligned outside a set of small density in C_k. By the Koebe distortion theorem, μ and $h_k^*(\nu)$ are also nearly aligned outside a set of small density in B_k. But $h_k^*(\nu)$ is uniformly close to $h_\infty^*(\nu)$, so in the limit $h_\infty^*(\nu) = \mu$ almost everywhere.

Thus μ is holomorphic on B_∞; indeed μ is dual to $h_\infty^*(dz^2)$. The proof is completed by the preceding lemma.

■

By the same reasoning used to deduce Corollary 3.10, we have:

Corollary 3.18 *If f is critically finite, then $J(f)$ carries no invariant line field, except when f is double covered by an integral torus endomorphism.*

This corollary also follows from the uniqueness part of Thurston's characterization of critically finite rational maps (Theorem B.2.)

More on the motivation for the no invariant line fields conjecture can be found in the expository article [Mc3]. The relation of this conjecture to the Teichmüller theory of a rational map is explained in [McS].

Chapter 4

Holomorphic motions and the Mandelbrot set

This chapter presents results about the Mandelbrot set and general holomorphic families of rational maps. In particular we explain the equivalence of the density of hyperbolic dynamics in the quadratic family and the absence of invariant line fields.

The idea of relating structural stability of rational maps to holomorphic motions of the Julia set is due to Mañé, Sad and Sullivan [MSS]; their methods form the basis of this chapter.

4.1 Stability of rational maps

Definitions. Let X be a connected complex manifold. A *holomorphic family of rational maps*, parameterized by X, is a holomorphic map $f : X \times \widehat{\mathbb{C}} \to \widehat{\mathbb{C}}$. We denote this map by $f_\lambda(z)$, where $\lambda \in X$ and $z \in \widehat{\mathbb{C}}$; then $f_\lambda : \widehat{\mathbb{C}} \to \widehat{\mathbb{C}}$ is a rational map.

Let x be a basepoint in X. A *holomorphic motion* of a set $E \subset \widehat{\mathbb{C}}$ parameterized by (X, x) is a family of injections

$$\phi_\lambda : E \to \widehat{\mathbb{C}},$$

one for each λ in X, such that $\phi_\lambda(e)$ is a holomorphic function of λ for each fixed e, and $\phi_x = \mathrm{id}$.

A basic fact about holomorphic motions is:

Theorem 4.1 (The λ-Lemma) *A holomorphic motion of E has a unique extension to a holomorphic motion of $\overline{E}$. The extended motion gives a continuous map $\phi : X \times \overline{E} \to \widehat{\mathbb{C}}$. For each λ, the map $\phi_\lambda : E \to \widehat{\mathbb{C}}$ extends to a quasiconformal map of the sphere to itself.*

See [MSS], [BR] and [ST] for details and further results.

Given a holomorphic family of rational maps f_λ, we say the corresponding Julia sets $J_\lambda \subset \widehat{\mathbb{C}}$ *move holomorphically* if there is a holomorphic motion

$$\phi_\lambda : J_x \to \widehat{\mathbb{C}}$$

such that $\phi_\lambda(J_x) = J_\lambda$ and

$$\phi_\lambda \circ f_x(z) \;=\; f_\lambda \circ \phi_\lambda(z)$$

for all z in J_x. Thus ϕ_λ provides a conjugacy between f_x and f_λ on their respective Julia sets. The motion ϕ_λ is unique if it exists, by density of periodic cycles in J_x.

The Julia sets move holomorphically *at x* if they move holomorphically on some neighborhood U of x in X.

A periodic point z of f_x of period n is *persistently indifferent* if there is a neighborhood U of x and a holomorphic map $w : U \to \widehat{\mathbb{C}}$ such that $w(x) = z$, $f_\lambda^n(w(\lambda)) = w(\lambda)$, and $|(f_\lambda^n)'(w(\lambda))| = 1$ for all λ in U. (Here $(f_\lambda^n)'(z) = df_\lambda^n/dz$.)

Theorem 4.2 (Characterizations of stability) *Let f_λ be a holomorphic family of rational maps parameterized by X, and let x be a point in X. Then the following conditions are equivalent:*

1. *The number of attracting cycles of f_λ is locally constant at x.*

2. *The maximum period of an attracting cycle of f_λ is locally bounded at x.*

3. *The Julia set moves holomorphically at x.*

4. *For all y sufficiently close to x, every periodic point of f_y is attracting, repelling or persistently indifferent.*

5. *The Julia set J_λ depends continuously on λ (in the Hausdorff topology) on a neighborhood of x.*

Suppose in addition that $c_i : X \to \widehat{\mathbb{C}}$, are holomorphic maps parameterizing the critical points of f_λ. Then the following conditions are also equivalent to those above:

6. For each i, the functions $\lambda \mapsto f_\lambda^n(c_i(\lambda))$, $n = 0, 1, 2, \ldots$ form a normal family at x.

7. There is a neighborhood U of x such that for all λ in U, $c_i(\lambda) \in J_\lambda$ if and only if $c_i(x) \in J_x$.

Definition. The open set $X^{\text{stable}} \subset X$ where any of the above equivalent conditions are satisfied is called the set of *J-stable parameters* of the family f_λ.

Proof. An attracting periodic cycle of f_x remains attracting, and of the same period, under a small change in x. Thus $1 \implies 2$.

We now show $2 \implies 3$. Assume the period of every attracting cycle is bounded by N on a polydisk neighborhood U of x. Then the repelling periodic points of period greater than N remain repelling throughout U; in particular, a repelling point cannot become indifferent because it must then become attracting nearby. But whenever two periodic points collide, the result is a multiple root of $f_\lambda^M(z) = z$, which is necessarily an indifferent periodic point (because the graph of $z' = f_\lambda^M(z)$ is tangent to the diagonal $z' = z$). Thus the repelling periodic points of sufficiently high period move holomorphically and without collision as λ varies in U. Since the repelling points of period greater than N are dense in the Julia set, the Julia set moves holomorphically by the λ-lemma.

To see $3 \implies 4$, note that if the Julia set moves holomorphically at x, then it moves holomorphically at y for all y sufficiently close to x. So it suffices to show that when the Julia set moves holomorphically at y, say by a motion $\phi : U \times J_y \to \mathbb{C}$ defined on a neighborhood U of y, then any indifferent periodic point z of period n for f_y is persistently indifferent.

If z lies outside the Julia set J_y, then it is not a parabolic point; in particular, $(f_x^n)'(z) \neq 1$ so we can locally parameterize this periodic point by an analytic function $w(\lambda)$, using the implicit function theorem. (In this case the graph of $z' = f_x^n(z)$ is transverse to the diagonal $z' = z$.) Since the Julia set moves continuously, $w(\lambda)$ remains outside J_λ for λ near y. Thus $|(f_\lambda^n)'(w(\lambda))| \leq 1$, so the derivative is constant and z is persistently indifferent. On the other hand, if z lies in J_y, then we may take $w(\lambda) = \phi_\lambda(z) \in J_\lambda$. Now $w(\lambda) \in J_\lambda$, so $|(f_\lambda^n)'(w(\lambda))| \geq 1$, and thus the derivative is again constant and z is persistently indifferent in this case as well.

Next we show $4 \implies 1$. Suppose there is a neighborhood U of x such that for every y in U, every indifferent periodic point of f_y is persistently indifferent. Then a periodic point cannot change from attracting to repelling over U (since it would have to pass through a non-persistent indifferent cycle). Thus the number of attracting cycles is locally constant at x. This shows $4 \implies 1$, and thus 1–4 are equivalent.

By the λ-lemma, $3 \implies 5$; to establish the equivalence of 5 with 1–4, it suffices to show $5 \implies 1$. So suppose J_λ varies continuously in the Hausdorff topology on a connected neighborhood U of x. By Siegel's theorem [Sie], [Bea2, Theorem 6.6.4], there is a dense subset $E \subset S^1$ such that any periodic cycle whose multiplier lies in E is the center of a Siegel disk. Since the center of a Siegel disk of f_λ lies a definite distance from J_λ, its multiplier cannot become repelling under a small perturbation; thus any cycle whose multiplier lies in E is persistently indifferent. Therefore the multiplier of an attracting or repelling cycle of f_λ cannot cross the unit circle as λ varies, and hence the number of attracting cycles of f_λ is constant on U.

To conclude, we treat the cases where the critical points of f_λ are parameterized by functions $< c_i(\lambda) : i = 1, \ldots, 2d - 2 >$.

We will first show $6 \implies 2$. Suppose the forward orbits of the critical points form normal families in λ on a polydisk neighborhood U of x. Let $g_i : U \to \hat{\mathbb{C}}$ be a holomorphic function obtained as the limit of a subsequence of $f_\lambda^n(c_i(\lambda))$ as $n \to \infty$. Suppose f_y has an attracting cycle of period N for some y in U. Since an attracting cycle attracts a critical point, the cycle includes a point of the form $g_i(y)$, and thus

$$f_\lambda^N(g_i(\lambda)) = g_i(\lambda)$$

when $\lambda = y$. This cycle remains attracting under a small change in λ, so the relation above holds on a neighborhood of y and thus for all λ in U. Therefore an attracting cycle which attracts the ith critical point has period at most N. Since there are only a finite number of critical points, we obtain an upper bound on the periods of attracting cycles which holds throughout U. Thus $6 \implies 2$.

Similarly, $7 \implies 2$. To see this, suppose there is a polydisk neighborhood U of x such that $c_i(\lambda) \in J_\lambda$ if and only if $c_i(x) \in J_\lambda$. Shrinking U if necessary, we can find three holomorphically varying points $z_j(\lambda)$, $j = 1, 2, 3$ such that $z_j(\lambda) \in J_\lambda$ for all λ in U; for example, $z_j(\lambda)$ can be chosen as a repelling periodic point for f_λ. Now suppose $y \in U$ and f_y has an attracting cycle that attracts $c_i(y)$. Then $c_i(y) \notin J_y$, so $c_i(\lambda)$ lies outside the Julia set J_λ for all λ in U. In particular the graphs of $f_\lambda^n(c_i(\lambda))$ and $z_j(\lambda)$ are disjoint over U, so by Montel's theorem (§2.5) the forward orbit of the ith critical point forms a normal family. Reasoning as above, we obtain a bound on the period of any attracting cycle that attracts $c_i(\lambda)$. Since there are only finitely many critical points, $7 \implies 2$.

Finally we show $3 \implies 6$ and 7. Suppose the Julia set moves by a holomorphic motion $\phi : U \times J_x \to \widehat{\mathbb{C}}$ defined on a neighborhood of x. Note that a point z in J_λ is a critical point of multiplicity m for f_λ if and only if the map $f_\lambda : J_\lambda \to J_\lambda$ is locally $(m+1)$-to-1 at z. (Here we use the fact that the Julia set is perfect and totally invariant).

Since $\phi_\lambda : J_x \to J_\lambda$ preserves the topological dynamics, it preserves the critical points, their multiplicities and their forward orbits. Thus $c_i(y) \in J_y$ for some y in U implies $c_i(\lambda) \in J_\lambda$ for all λ in U, and $\phi_\lambda(c_i(x)) = c_i(\lambda)$. Therefore $3 \implies 7$.

Now pick three points z_1, z_2 and z_3 in J_x which are disjoint from the forward orbits of the critical points of f_x. Then $\phi_\lambda(z_j)$ is disjoint from the forward orbits of the critical points of f_λ for all λ in U. By Montel's theorem, the forward orbits of the critical points form normal families on U, so $3 \implies 6$.

∎

Theorem 4.3 ([MSS]) *The set X^{stable} of J-stable parameters is an open dense subset of X.*

Proof. Let $N(\lambda)$ denote the number of attracting periodic cycles of f_λ. Then $N(\lambda)$ is bounded above by $2d - 2$, where d is the degree of the rational maps in the family. Since attracting cycles persist under small changes in λ, we have $N(\lambda) \leq \limsup N(\lambda_n)$ whenever $\lambda_n \to \lambda$. Thus the set of local maxima of $N(\lambda)$ is open and dense, and these maxima coincide with the set of J-stable parameters, by Theorem 4.2, case 1.

∎

Definition. The $\lambda \in X$ such that f_λ is hyperbolic form the *hyperbolic parameters* X^{hype}.

Theorem 4.4 *In any holomorphic family of rational maps, the hyperbolic parameters form an open and closed subset of the J-stable parameters.*

Proof. The condition that all critical points tend to attracting cycles is clearly open, and it implies structural stability by Theorem 4.2, case 7. Thus the hyperbolic parameters are an open subset of X^{stable}. On the other hand, if f_λ is structurally stable, any critical point or parabolic cycle in $J(f_\lambda)$ persists under small changes in λ; thus the non-hyperbolic structurally stable parameters also form an open set.

∎

Definition. A rational map f of degree d is *J-structurally stable* if it is J-stable in the family of all rational maps of degree d.

By Theorems 4.3 and 4.4, the density of hyperbolic dynamics within the space of all rational maps of degree d (Conjecture 1.1) is equivalent to:

Conjecture 4.5 *A J-structurally stable rational map of degree d is hyperbolic.*

What did Fatou conjecture? In his second memoir, Fatou touches on this circle of ideas. In his notation, R is a rational map, $\mathcal{F} = J(R)$ and $E_c + E'_c = P(f)$ (the forward orbit of the critical points union their limit points). Speaking of hyperbolicity, he writes [Fatou2, p.73]:

> Il est probable, mais je n'ai pas approfondi la question, que cette propriété appartient à toutes les substitutions générales, c'est-à-dire celles dont les coefficients ne vérifient aucune relation particulière. Je signale, dans ce même ordre d'idées, l'intérèt qu'il y aurait à rechercher les conditions nécessaires et suffisantes pour que l'ensemble $\mathcal{F}$ varie d'une manière continue, tant au point de vue de la position de ses points qu'au point de vue de la connexion des domains dans lesquels il divise le plan, lorsqu'on fait varier les coefficients de $R(z)$. Il paraît bien et l'on peut le constater sur des exemples que la discontinuité a lieu pour les valeurs des coefficients, telles que $\mathcal{F}$ contienne des points de $E_c + E'_c$.

Thus for Fatou hyperbolic dynamics is probably dense, although his first sentence may mean that the non-hyperbolic rational maps should be contained in a countable union of proper subvarieties. (This is false, by an elementary argument [Lyu2, Proposition 3.4]; in fact, the non-hyperbolic rational maps of any given degree have positive measure [Rees2].) Fatou also states that the Julia set $J(f)$ appears to vary discontinuously exactly at the parameters where it meets $P(f)$. Since the former condition is equivalent to failure of J-structural stability, and the latter is equivalent to failure of hyperbolicity, in hindsight we can interpret Fatou's observation as a version of Conjecture 4.5.[1]

4.2   The Mandelbrot set

We now specialize to the family of quadratic polynomials $f_c(z) = z^2 + c$ for $c \in X = \mathbb{C}$.

The *Mandelbrot set* is defined by

$$M = \{c : f_c^n(0) \text{ does not tend to } \infty \text{ as } n \to \infty\}.$$

---

[1]This reference and its discussion were contributed by Eremenko, Lyubich and Milnor.

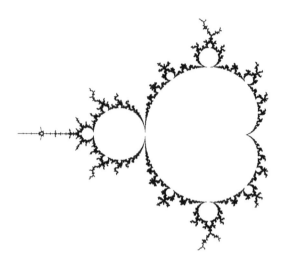

Figure 4.1. The boundary of the Mandelbrot set.

Theorem 4.6 *The boundary of the Mandelbrot set is the same as the set of c such that the functions $< c \mapsto f_c^n(0) : n = 1, 2, 3, \ldots >$ do not form a normal family near c. Thus $X^{stable} = \mathbb{C} - \partial M$, where X^{stable} denotes the set of J-stable parameters of the family f_c.*

Proof. There exists an R such that once the forward orbit of the critical point leaves the ball of radius R about the origin, it tends to infinity. (In fact one may take $R = 2$.) Thus the forward orbit of the critical point $z = 0$ is bounded by R for all c in the interior of M. Outside of M the critical point tends to infinity locally uniformly, so in either case $< f_c^n(0) >$ forms a normal family. For c on the boundary of M, $f_c^n(0)$ is bounded by R while $f_{c'}^n(0) \to \infty$ for nearby values c', so $< f_c^n(0) >$ is not normal on at any point of ∂M. Thus $\mathbb{C} - \partial M$ is exactly the domain of normality of the forward orbit of the critical point. By Theorem 4.2, this set coincides with the set of J-stable parameters. (The critical point at infinity is fixed so its iterates form a normal family for all values of c).

∎

Theorem 4.7 *For c in the Mandelbrot set, $f_c(z) = z^2 + c$ is hyperbolic if and only if f_c has an attracting cycle in $\mathbb{C}$.*

Proof. If f_c is hyperbolic and $c \in M$, then the critical point $z = 0$ tends to an attracting cycle, which must lie in $\mathbb{C}$ since the forward orbit of the critical point is bounded.

Conversely, if f_c has a finite attracting cycle, this cycle must attract the critical point $z = 0$, so $c \in M$; and f_c is hyperbolic because its other critical point $z = \infty$ is already a superattracting fixed point.

∎

Definition. A component U of the interior of the Mandelbrot set M is *hyperbolic* if f_c is hyperbolic for some c in U. By Theorem 4.4, if U is hyperbolic, then f_c is hyperbolic for all c in U.

Theorem 4.8 *If f_c has an indifferent cycle, then c lies in the boundary of the Mandelbrot set.*

Proof. By Theorem 4.2, if f_c has an indifferent periodic point for c in $X^{\text{stable}} = \mathbb{C} - \partial M$, then this point is persistently indifferent. But then f_c has an indifferent cycle for every c, contrary to the fact that $f_0(z) = z^2$ has no such cycle.

∎

Theorem 4.9 (Line fields and hyperbolicity) *A point c belongs to a non-hyperbolic component U of the interior of the Mandelbrot set if and only if the Julia set $J(f_c)$ has positive measure and carries an invariant line field.*

Proof. Suppose c belongs to a non-hyperbolic component U of the interior of M. Then f_c has no attracting or indifferent cycles. A polynomial has no Herman rings, so by the classification of periodic components, the Fatou set of f_c consists solely of the basin of attraction of $z = \infty$. Consequently the Julia set $J(f_c)$ is full (it does not disconnect the plane.)

For $\lambda \in U$, let

$$\phi_\lambda : (\mathbb{C} - K(f_c)) \to (\mathbb{C} - K(f_\lambda))$$

be the unique holomorphic conjugacy between f_c and f_λ on their basins of infinity. (Compare Theorem 3.3.) The map $\phi_\lambda(z)$ varies holomorphically in both λ and z; indeed

$$\phi_\lambda(z) \quad = \quad \lim f_\lambda^{-n} \circ f_c^n(z)$$

for appropriate branches of the inverse. Thus ϕ_λ defines a holomorphic motion of $\mathbb{C} - J(f_c)$.

By the λ-lemma (Theorem 4.1), this motion extends to a motion of the closure of the basin of infinity, which is equal to $\widehat{\mathbb{C}}$ because $J(f_c)$ is full. Let $\overline{\phi}_\lambda : \widehat{\mathbb{C}} \to \widehat{\mathbb{C}}$ be this extended motion; for each fixed λ it is a quasiconformal map, which is conformal outside of $J(f_c)$.

If $\overline{\phi}_\lambda$ is conformal on the whole sphere, then f_c and f_λ are conformally conjugate, which is impossible unless $c = \lambda$. Thus for $\lambda \neq c$ in U the complex dilatation

$$\mu_\lambda \quad = \quad \frac{\partial_z \overline{\phi}_\lambda}{\partial_z \overline{\phi}_z}$$

is supported on a subset of $J(f_c)$ of positive measure. Moreover $f_c^*(\mu_\lambda) = \mu_\lambda$ because ϕ_λ is a conjugacy between f_c and f_λ. Thus $\mu = \mu_\lambda/|\mu_\lambda|$ defines an invariant line field for f_c.

Conversely, suppose $J(f_c)$ supports an invariant line field μ. By the "measurable Riemann mapping theorem" (§2.6) there is a holomorphic family of quasiconformal maps $\phi_t : \mathbb{C} \to \mathbb{C}$ with complex dilatation

$$\frac{\overline{\partial}_z \phi_t}{\partial_z \phi_t} \quad = \quad t\mu$$

for all t in the unit disk Δ. By invariance of μ under f_c, the map

$$g_t(z) \quad = \quad \phi_t \circ f_c \circ \phi_t^{-1}(z)$$

is a quadratic polynomial depending holomorphic on t (cf. Theorem 2.13). With suitable normalizations we can assume $\phi_t'(\infty) = 1$ and $g_t(z) = z^2 + c(t)$.

We claim $c(t)$ is an injective function of t. Indeed, if $c(t_1) = c(t_2)$, then the Julia sets of g_{t_1} and g_{t_2} are the same. Since ϕ_t is conformal outside the Julia set and normalized at infinity, it follows that $\phi_{t_1} = \phi_{t_2}$ on $\mathbb{C} - K(f_c)$. By continuity these maps agree on the Julia set $J(f_c)$, and therefore on the whole plane. Thus their complex dilatations are the same, i.e. $t_1\mu = t_2\mu$. Since μ is not zero, $t_1 = t_2$.

Next observe that $z^2 + c(t)$ is topologically conjugate to $z^2 + c(0)$; since $c = c(0)$ lies in the Mandelbrot set, so does $c(t)$. By injectivity, the image of the unit disk under $c(t)$ gives an open subset of M containing c. Thus c lies in a component U of the interior of the Mandelbrot set. Since the Julia set has positive measure, f_c is not hyperbolic, so neither is the corresponding component U of the interior of M.

∎

Corollary 4.10 *Hyperbolic dynamics is dense in the quadratic family if and only if there is no quadratic polynomial with an invariant line field on its Julia set.*

Proof. Clearly f_c is hyperbolic for c outside the Mandelbrot set, since the critical point $z = 0$ is attracted to the superattracting fixed point at infinity. Points c in the Mandelbrot set are handled by the preceding theorem.

∎

This shows Conjectures 1.2 and 1.5 are equivalent. A stronger form of Conjecture 1.2 is:

Conjecture 4.11 *The boundary of the Mandelbrot set is locally connected.*

It has been shown by Douady and Hubbard that Conjecture 4.11 implies Conjecture 1.2 [DH1, Exposé XXII]. The methods of Yoccoz also yield results about local connectivity of M; see [Mil3], [Yoc], and §8.

Chapter 5

Compactness in holomorphic dynamics

Let f be a rational map with an invariant line field μ on its Julia set. In §3 we analyzed what happens when a point x of almost continuity of μ does not tend to the postcritical set under iteration. Using the dynamics, we found one can expand small neighborhoods of x and pass to a limit to obtain a holomorphic invariant line field defined on a ball of definite size.

In the sequel we will consider what happens if the point of almost continuity *does* tend to the postcritical set. In this case, the expansion in the Poincaré metric on $\widehat{\mathbb{C}} - P(f)$ guaranteed by Theorem 3.6 allows us to at least obtain a nearly holomorphic line field defined on a Poincaré ball of definite size. Unfortunately, the spherical diameter of such a ball may tend to zero.

If the mapping is renormalizable, there is still a chance of rescaling high iterates of f near the postcritical set to obtain a limiting dynamical system with a holomorphic invariant line field. The presence of a critical point makes this impossible (Theorem 5.13). This contradiction rules out the presence of an invariant line field for the original dynamical system f.

To obtain a limit of the rescaled dynamical systems, we need to go beyond the setting of iterated rational maps. For our purposes, it will suffice to construct a limiting *proper map*; under good conditions, the limit will also also be *polynomial-like* in the sense of Douady and Hubbard.

65

In this chapter we develop compactness results for line fields and dynamical systems to carry through the argument above.

5.1   Convergence of Riemann mappings

Definitions. A *disk* is an open simply-connected region in $\mathbb{C}$, possibly equal to $\mathbb{C}$ itself.

Consider the set $\mathcal{D}$ of pointed disks (U, u). The *Carathéodory topology* on $\mathcal{D}$ is defined as follows: $(U_n, u_n) \to (U, u)$ if and only if

(i) $u_n \to u$;

(ii) for any compact $K \subset U$, $K \subset U_n$ for all n sufficiently large; and

(iii) for any open *connected* N containing u, if $N \subset U_n$ for infinitely many n, then $N \subset U$.

Equivalently, convergence means $u_n \to u$, and for any subsequence such that $(\widehat{\mathbb{C}} - U_n) \to K$ in the Hausdorff topology on compact sets of the sphere, U is equal to the component of $\widehat{\mathbb{C}} - K$ containing u.

Note that pieces of U_n can "pinch off" and disappear in the limit; for example, if U_n is the union of a unit disk centered at zero and another centered at $1 + 1/n$, then $(U_n, 0)$ converges to $(\Delta, 0)$.

Let $\mathcal{E} \subset \mathcal{D}$ denote the subspace of disks not equal to $\mathbb{C}$.

Let $\mathcal{F}$ be the space of univalent maps $f : \Delta \to \mathbb{C}$ such that $f'(0) > 0$, equipped with the topology of uniform convergence on compact sets. There is a natural bijection $\mathcal{E} \to \mathcal{F}$ which associates to each (U, u) the unique Riemann mapping $f : (\Delta, 0) \to (U, u)$ such that $f'(0) > 0$.

Theorem 5.1 *The natural map $\mathcal{E} \to \mathcal{F}$ from disks to Riemann mappings is a homeomorphism.*

See [Oes, §4], [Car1, §119-123].

The following facts are easily verified:

Theorem 5.2 *The set of disks $(U, 0)$ containing $B(0, r)$ for some $r > 0$ is compact in $\mathcal{D}$.*

Theorem 5.3 *If $(U_n, u_n) \to (U, u)$ in $\mathcal{E}$, and the distance $d(u_n, w_n) < D$ in the hyperbolic metric on U_n, then there is a further subsequence such that $(U_n, w_n) \to (U, w)$.*

Carathéodory topology on functions. Let $\mathcal{H}$ be the set of all holomorphic functions $f : (U, u) \to \mathbb{C}$ defined on pointed disks $(U, u) \in \mathcal{D}$.

We define the Carathéodory topology on $\mathcal{H}$ as follows.

Let $f_n : (U_n, u_n) \to \mathbb{C}$ be a sequence in $\mathcal{H}$. Then f_n converges to $f : (U, u) \to \mathbb{C}$ if:

(i) $(U_n, u_n) \to (U, u)$ in $\mathcal{D}$, and

(ii) for all n sufficiently large, f_n converges to f uniformly on compact subsets of U.

Any compact set $K \subset U$ is eventually contained in U_n, so f_n is defined on K for all n sufficiently large.

For example, one may easily check:

Theorem 5.4 *If $(U_n, u_n) \to (U, u)$ in $\mathcal{E}$, then $f_n^{-1} \to f^{-1}$, where f_n and f are the corresponding Riemann maps in $\mathcal{F}$.*

In the sequel, convergence of holomorphic functions will always be meant to take place in this topology.

5.2   Proper maps

Definitions. Let U, V be a pair of disks. A *proper map between disks* $f : U \to V$ is a holomorphic map such that $f^{-1}(K)$ is compact for every compact set $K \subset V$. Then $f^{-1}(x)$ is finite for all x in V, and the cardinality of the inverse image of a point (counted with multiplicity) is the *degree* of f. The *critical points* of f are denoted $C(f)$.

To employ the Carathéodory topology, it is useful to add base-points to the disks U and V. The notation $f : (U, u) \to (V, v)$ means $u \in U$, $v \in V$ and $f(u) = v$.

Lemma 5.5 *Let $f : U \to V$ be a proper map of degree d with critical values lying in a compact set $K \subset V$, and let $K' = f^{-1}(K)$. Then:*

1. $\mathrm{mod}(f^{-1}(A)) = \mathrm{mod}(A)/d$ *for any annulus* $A \subset V$ *enclosing* K.

2. $\mathrm{mod}(K',U) \geq \mathrm{mod}(K,V)/d$.

3. *If* $U \neq \mathbb{C}$, *then* $\mathrm{diam}(K') \leq D(\mathrm{mod}(K,V))$ *in the hyperbolic metric on* U, *where* $D(m) \to 0$ *as* $m \to \infty$.

Proof. Since a proper local homeomorphism is a covering map, $f^{-1}(A)$ is an annulus covering A by degree d; this gives the first claim. An annulus separating K from ∂V has a preimage separating K' from ∂U, so the second claim follows from the first. The third claim follows from Theorem 2.4.

∎

Theorem 5.6 (Limits of proper maps) *Let* (U_n, u_n) *and* (V_n, v_n) *be a sequence of disks converging to* (U, u) *and* (V, v) *respectively. Let* $f_n : (U_n, u_n) \to (V_n, v_n)$ *be a sequence of proper maps of degree* d. *Then after passing to a subsequence, either*

1. $U = \mathbb{C}$ *and* f_n *converges to the constant map* $f(z) = v$; *or*

2. $V = \mathbb{C}$ *and* $f_n(x) \to \infty$ *for every* $x \in U$ *with at most* d *exceptions; or*

3. f_n *converges to* $f : (U, u) \to (V, v)$, *a proper map of degree less than or equal to* d.

In the last case, if there is a compact K *such that the critical points* $C(f_n) \subset K \subset U$ *for all* n *sufficiently large, then the limit* f *has degree* d.

A more precise statement of case 2 is the following: there is a set $E \subset U$ with $|E| \leq d$ such that for all $x \in U$ and for all n sufficiently large, $x \in U_n$ (by the definition of Carathéodory convergence) and $f_n(x) \to \infty$.

Proof. The proof will be broken into 3 cases: (I) $V = \mathbb{C}$, (II) $V \neq \mathbb{C}$ but $U = \mathbb{C}$, and (III) neither U nor $V = \mathbb{C}$.

I. $V = \mathbb{C}$. Since f_n has at most d critical values, we can choose $R > 0$ and pass to a subsequence such that every critical value of f_n either lies in the Euclidean ball $B(v, R)$ or tends to infinity as $n \to \infty$.

Suppose case 2 of the Theorem does not hold. Then (after passing to a subsequence and possibly increasing R) we can assume there is a set $E \subset U$ with $|E| = d + 1$ and $f_n(E) \subset B(v, R)$ for all n.

Consider the annulus $A(S) = B(v, S) - B(v, R)$ for $S > R$. For all n sufficiently large, f_n has no critical values in $A(S)$, so $f_n^{-1}(A(S))$ consists of at most d annuli each mapping to $A(S)$ by a covering map of degree at most d. The union of these annuli separate E from ∞, so there is a component $B_n(S)$ of $f_n^{-1}(A(S))$ which separates a two point set $\{e_1, e_2\} \subset E$ from ∞ (using the fact that $|E| = d + 1$). After passing to a further subsequence, we can assume the same two points e_1, e_2 work for all n.

Since $B_n(S)$ is a covering of $A_n(S)$ with degree at most d, we have $\mod B_n(S) \geq \mod A_n(S)/d$. By Theorem 2.1, when S is large $B_n(S)$ contains a round annulus $R_n(S)$ of modulus at least $\mod(A(S))/d - O(1)$. Since $R_n(S)$ encloses the set $\{e_1, e_2\}$, its outer boundary is a circle of diameter at least $C|S|^{1/d}$ for some C depending only on $|e_1 - e_2|$ and d. Therefore $U = \mathbb{C}$, and there is a constant C' such that for any compact set $L \subset \mathbb{C}$, $|f_n(z)| \leq C'(1 + |z|^d)$ for z in L and all n sufficiently large. By this estimate, after passing to a subsequence, f_n converges to a polynomial f of degree at most d. If f is constant, then case 1 holds, otherwise we are in case 3.

To finish, we check the last statement of the theorem. That is, suppose we are in case 3 and the critical points $C(f_n)$ lie in a compact set K for all n. Then $f_n'(z)$ has $d - 1$ zeros in K for all n, so the limiting polynomial f is also of degree d.

II. $V \neq \mathbb{C}$ but $U = \mathbb{C}$. Then the Schwarz lemma shows f_n converges to the constant function v.

III. Neither U nor V is equal to $\mathbb{C}$. Then for n large enough, U_n and V_n are also different from $\mathbb{C}$. Let $\alpha_n : (\Delta, 0) \to (U_n, u_n)$ and $\beta_n : (\Delta, 0) \to (V_n, v_n)$ be the unique Riemann mappings with positive derivatives at the origin. By Theorem 5.1, these maps converge to Riemann mappings $\alpha : (\Delta, 0) \to (U, u)$ and $\beta : (\Delta, 0) \to (V, v)$ respectively.

There is a unique proper map F_n of degree d such that the diagram

$$
\begin{array}{ccc}
(\Delta, 0) & \xrightarrow{\ F_n\ } & (\Delta, 0) \\
\alpha_n \downarrow & & \beta_n \downarrow \\
(U_n, u_n) & \xrightarrow{\ f_n\ } & (V_n, v_n)
\end{array}
$$

commutes. Then F_n can be written as a Blaschke product

$$
F_n(z) \;=\; e^{i\theta_n} z \prod_{1}^{d-1} \frac{z - a_i(n)}{1 - \overline{a_i(n)}z},
$$

where 0 and $a_i(n) \in \Delta$ are the preimages of 0. After passing to a subsequence, we can assume θ_n and $a_i(n)$ converge, so F_n converges to a proper map F of degree between 1 and d. (The degree is less than d if and only if $|a_i(n)| \to 1$ for some i.) It follows that f_n converges to $f = \beta \circ F \circ \alpha^{-1}$, so we are in case 3.

Finally we verify the last statement of the theorem in this case as well. If the critical values of f_n lie in a compact set $K \subset V$, then the critical values of F_n lie within a compact set L, $0 \in L \subset \Delta$, for all n sufficiently large. By Lemma 5.5, $L' = F_n^{-1}(L)$ has bounded hyperbolic diameter, and contains 0 as well as $\{a_1(n), \ldots, a_{d-1}(n)\}$. Thus for each i, $a_i(n)$ tends to a limit in the open unit disk, the limit F has degree d and therefore f has degree d.

■

Example. Let $f_n : \mathbb{C} \to \mathbb{C}$ be a sequence of polynomials of degree d with $f_n(0) = 0$. If the coefficients of f_n are bounded, then there is a subsequence converging to a polynomial of degree at most d. If the coefficients are unbounded, we can write $f_n = \alpha_n g_n$ for scalars $\alpha_n \to \infty$ and polynomials g_n with bounded coefficients, at least one of which has modulus one. Passing to a subsequence we have $g_n \to g$, where g is a nonconstant polynomial of degree at most d. (The limit is nonconstant because $f_n(0) = 0$ implies the constant coefficient of g_n is zero). Then $f_n(z) \to \infty$ for all z which are not among the zeros of g. The zeros of g determine the exceptional set in case 2 of Theorem 5.6.

5.3   Polynomial-like maps

A rational map may have a restriction which behaves like a polynomial, sometimes of much lower degree. The simplest example comes from an attracting or repelling fixed point, near which the map behaves like a polynomial of degree one. To capture behavior of higher degrees, Douady and Hubbard introduced the idea of a polynomial-like map [DH2].

First consider a polynomial $f : \mathbb{C} \to \mathbb{C}$ of degree $d > 1$. When $|z|$ is large, the behavior of f is dominated by its leading coefficient, so $|f(z)| \asymp |z|^d$. Thus for any sufficiently large disk $V = \{z \; : \; |z| < R\}$, the preimage $U = f^{-1}(V)$ is a smaller disk with compact closure in V. By definition, the filled Julia set $K(f)$ is the set of all z for which $f^n(z)$ remains bounded as $n \to \infty$; therefore $K(f) = \bigcup_{n>0} f^{-n}(V)$.

We now turn to the notion of a polynomial-like map, which abstracts the properties of the restriction of a polynomial to a large disk.

Definitions. A *polynomial-like map* $f : U \to V$ is a proper map between disks such that $\overline{U}$ is a compact subset of V. (It follows that neither U nor V is equal to $\mathbb{C}$).

The *filled Julia set* $K(f)$ is defined by

$$K(f) = \bigcap_1^\infty f^{-n}(V).$$

It is easy to see that $K(f)$ is *full* (it does not disconnect the plane).

The *Julia set* $J(f)$ is equal to the boundary of $K(f)$ in $\mathbb{C}$. The *postcritical set* $P(f) \subset V$ is defined as the closure of the forward orbits of the critical points of f.

Two polynomial-like maps f and g are *hybrid equivalent* if there is a quasiconformal conjugacy ϕ between f and g, defined on a neighborhood of their respective filled Julia sets, such that $\overline{\partial}\phi = 0$ on $K(f)$ (see [DH2, p.296]).

Theorem 5.7 *Every polynomial-like map f is hybrid equivalent to (a suitable restriction of) a polynomial g of the same degree. When $K(f)$ is connected, the polynomial g is unique up to affine conjugation.*

See [DH2, Theorem 1]. It follows, for example, that repelling periodic points are dense in the Julia set of f, and:

> $K(f)$ is connected if and only if it contains every critical point of f.

So when $K(f)$ is connected, $P(f) \subset K(f)$.

The Douady-Hubbard definition of a polynomial-like map does not include polynomials as a special case. It is often useful to adjoin polynomials to the maps under consideration, as in the compactness result below.

Definitions. Let $Poly_d$ denote the space of polynomial-like maps $f : (U, u) \to (V, v)$ and polynomials $f : (\mathbb{C}, u) \to (\mathbb{C}, v)$ of degree d, with connected Julia sets and basepoints $u \in K(f)$. We give $Poly_d$ the Carathéodory topology.

The space $Poly_d(m) \subset Poly_d$ consists of all polynomials of degree d and all polynomial-like maps with $\mod(U, V) \geq m > 0$.

Theorem 5.8 *The space $Poly_d(m)$ is compact up to affine conjugation.*

More precisely, any sequence $f_n : (U_n, u_n) \to (V_n, v_n)$ in $Poly_d(m)$, normalized so $u_n = 0$ and so the Euclidean diameter of $K(f_n)$ is equal to 1, has a convergent subsequence.

Proof. Assume $f_n : (U_n, 0) \to (V_n, v_n)$ in $Poly_d(m)$ is normalized so $\operatorname{diam}(K(f_n)) = 1$. Then $|v_n| \leq 1$ since 0 and $f_n(0) = v_n$ are in $K(f_n)$. By assumption $\mod(U_n, V_n) \geq m$, so $\mod(V_n - K(f_n)) > m$ and $\mod(U_n - K(f_n)) > m/d$, since the second annulus is a degree d cover of the first. By Theorem 2.5, the Euclidean distance from 0 to ∂U_n and from v_n to ∂V_n is greater than $C(m) > 0$. Thus by Theorem 5.2, we can pass to a subsequence such that $(U_n, 0) \to (U, 0)$ and $(V_n, v_n) \to (V, v)$.

Suppose $U = \mathbb{C}$; then $V = \mathbb{C}$ since $U \subset V$. Since f_n maps its filled Julia set of diameter one to itself, there is no subsequence such that f_n converges to a constant map, nor can f_n tend to infinity on $U - E$ where E is a finite set. So by Theorem 5.6 (Limits of proper maps), there is a further subsequence such that $f_n \to f$, a polynomial of degree between 1 and d. But the critical points of f_n

all lie in $K(f_n)$, and therefore within distance one of the origin, so the same theorem shows the limit is a polynomial of degree d. Thus $f \in Poly_d(m)$.

Now suppose $U \neq \mathbb{C}$. Then $\mathrm{mod}(U_n - K(f_n)) < M$ for all n, so $\mathrm{mod}(V_n - K(f_n)) < dM$. It follows that $V \neq \mathbb{C}$ as well. By Theorem 5.6(Limits of proper maps), there is a further subsequence such that $f_n \to f : (U, 0) \to (V, v)$. The limiting map f is proper with $1 \le \deg(f) \le d$.

The lower bound $\mathrm{mod}(U_n - K(f_n)) > m/d$ provides an upper bound on the diameter of $K(f_n)$ in the hyperbolic metric on U_n, by Theorem 2.4. By Theorem 5.3, any sequence $k_n \in K(f_n)$ eventually lies in a compact subset of U. It follows that $C(f_n) \subset K \subset U$ for a compact set K and all n sufficiently large, so f has degree d. By similar reasoning, the critical points do not escape from U under iteration, nor does the basepoint 0.

To check $\mathrm{mod}(U, V) \ge m$, let $h_n : A(2\pi \exp(m)) \to V_n - \overline{U_n}$ be a univalent map of a standard annulus of modulus m into the annular region between U_n and V_n. Then one can extract a limiting injection into $V - \overline{U}$, using Corollary 2.8.

In particular, $\overline{U}$ is a compact subset of V. Thus f is a polynomial-like map. Since the critical points of f do not escape under iteration, $K(f)$ is connected. Therefore $f \in Poly_d(m)$.

$\blacksquare$

Theorem 5.9 *The function $f \mapsto \mathrm{diam}(K(f))$ is continuous on $Poly_d$.*

Proof. Suppose $f_n \to f$ in $Poly_d$. One can form a restricted polynomial-like $f : U' \to V'$ with V' arbitrarily close to $K(f)$. Then $f_n : f_n^{-1}(V') \to V'$ is polynomial-like of degree d for all n sufficiently large, so $K(f_n)$ is eventually contained in a small neighborhood V' of $K(f)$. Thus $\limsup \mathrm{diam}(K(f_n)) \le \mathrm{diam}(K(f))$.

On the other hand, there are repelling periodic points x_1, x_2 in $J(f)$ with $d(x_1, x_2)$ arbitrarily close to $\mathrm{diam}(K(f))$. These persist under small perturbations, establishing the opposite inequality $\liminf \mathrm{diam}(K(f_n)) \ge \mathrm{diam}(K(f))$.

$\blacksquare$

Corollary 5.10 *If* $f \in \mathcal{P}oly_d(m)$ *has no attracting fixed point in* $\mathbb{C}$, *then*

$$\operatorname{diam} K(f) \;\leq\; C(d,m)\operatorname{diam} P(f)$$

in the Euclidean metric.

Proof. If not, we can find a sequence $f_n \in \mathcal{P}oly_d(m)$ such that

$$\frac{\operatorname{diam}(P(f_n))}{\operatorname{diam}(K(f_n))} \to 0.$$

By the compactness result above, after rescaling and passing to a subsequence we can assume f_n converges to a polynomial-like map $f : (U, u) \to (V, v)$ of degree d. Since $\operatorname{diam}(K(f_n)) \to \operatorname{diam}(K(f))$, we have $\operatorname{diam}(P(f_n)) \to 0$. It follows that $|P(f)| = 1$, so f has a superattracting fixed point. But then f_n has an attracting fixed point for all n sufficiently large.

∎

Another approach to the compactness of $\mathcal{P}oly_d(m)$ is via the theory of quasiconformal maps. Given a polynomial-like map $f : U \to V$, one can find a quasiconformal map $\phi : \mathbb{C} \to \mathbb{C}$ establishing a conjugacy between f and a polynomial near their respective filled Julia sets. A lower bound on $\operatorname{mod}(U,V)$ gives control on both the neighborhood of $K(f)$ where the conjugacy is defined, and on the dilatation $K(\phi)$. Then one can appeal to compactness results in the finite-dimensional space of polynomials, and compactness of quasiconformal maps with bounded dilatation.

5.4 Intersecting polynomial-like maps

The intersection of two polynomial-like maps is again polynomial-like, at least on each component of the intersection of the domains which maps over itself. This observation will prove useful in the sequel to establish coherence between various renormalizations of an iterated quadratic polynomial; it is made precise below.

Theorem 5.11 *Let $f_i : U_i \to V_i$ be polynomial-like maps of degree d_i, for $i = 1, 2$. Assume $f_1 = f_2 = f$ on $U = U_1 \cap U_2$. Let U' be a component of U with $U' \subset f(U') = V'$. Then*

$$f : U' \to V'$$

is polynomial-like of degree $d \leq \max(d_1, d_2)$, and

$$K(f) = K(f_1) \cap K(f_2) \cap U'.$$

If $d = d_i$, then $K(f) = K(f_i)$.

Proof. Let $V = V_1 \cap V_2$. We first remark that $f : U \to V$ is proper: this is immediate from the fact that

$$f^{-1}(E) = f_1^{-1}(E) \cap f_2^{-1}(E)$$

is compact if E is compact. Therefore $f : U' \to V'$ is proper and V' is a component of V. It is clear that U' and V' are disks since they are components of intersections of disks in $\mathbb{C}$. Finally $\overline{U'}$ is a compact subset of V' because $\overline{U}$ is a compact subset of V.

Thus f is polynomial-like. Its filled Julia set is given by

$$
\begin{aligned}
K(f) &= \bigcap f^{-n}(V') = \bigcap (f_1^{-n}(V_1) \cap f_2^{-n}(V_2) \cap U') \\
&= K(f_1) \cap K(f_2) \cap U'.
\end{aligned}
$$

A point in $K(f_i)$ has d_i preimages (counted with multiplicity) under f_i, and d under f; since the graph of f is contained in that of f_i, we have $d \leq d_i$.

If $d = d_i$, then $f^{-1}(x) = f_i^{-1}(x)$ for any $x \in K(f)$. The backward orbit of a point in the Julia set is dense in the Julia set, so $J(f) = \partial K(f) = J(f_i) = \partial K(f_i)$, and therefore $K(f) = K(f_i)$. $\blacksquare$

5.5   Polynomial-like maps inside proper maps

Let $f : U \to V$ be a proper map between disks.

We do not assume that $U \subset V$.

We will state a criterion allowing one to extract a polynomial-like map $f : U' \to V'$.

Definitions. The proper map f is *critically compact* if its critical points remain in U under forward iteration and the postcritical set

$$P(f) \;=\; \overline{\bigcup_{n>0, c \in C(f)} f^n(c)}$$

is a compact subset of U (and therefore of V).

Theorem 5.12 *Let $f : U \to V$ be a critically compact proper map of degree $d > 1$. There is a constant M_d such that when $\mathrm{mod}(P(f), V) > M_d$, either*

  1. *f has an attracting fixed point in U, or*

  2. *there is a restriction $f : U' \to V'$ which is a polynomial-like map of degree d with connected Julia set.*

Here $P(f) \subset U' \subset U$, and U' can be chosen so that

$$\mathrm{mod}(U', V') > m_d(\mathrm{mod}(P(f), V)) > 0$$

where $m_d(x) \to \infty$ as $x \to \infty$.

The two possibilities above are not exclusive.

Proof. Let $f_n : (U_n, 0) \to (V_n, v_n)$ be a sequence of critically compact proper maps of degree d, with no attracting fixed points and with $M_n = \mathrm{mod}(P(f_n), V_n) \to \infty$. Here we have normalized so that 0 is a critical point of f_n. It suffices to show that after passing to a subsequence f_n is polynomial-like of degree d for all n sufficiently large, and that the polynomial-like restriction $f_n : U'_n \to V'_n$ can be taken with $\mathrm{mod}(U'_n, V'_n) \to \infty$.

Let $Q(f_n) = f_n^{-1}(P(f_n)) \supset P(f_n)$. Then $0 \in Q(f_n)$ and $|Q(f_n)| > 1$, for otherwise $Q(f_n)$ would consist of a single superattracting fixed point for f_n. Further normalizing by scaling, we can assume that $\mathrm{diam}(Q(f_n)) = 1$ in the Euclidean metric. Then $|v_n| \leq 1$, so passing to a subsequence we can assume v_n converges to a point v in $\mathbb{C}$.

By Lemma 5.5, $\mathrm{mod}(Q(f_n), U_n) \geq M_n/d \to \infty$. Thus $(U_n, 0) \to (\mathbb{C}, 0)$ in the Carathéodory topology. Since f_n has no attracting fixed point, the Schwarz lemma implies $(V_n, v_n) \to (\mathbb{C}, v)$.

We claim that for a further subsequence, $f_n : (U_n, 0) \to (V_n, v_n)$ converges to a polynomial $g : \mathbb{C} \to \mathbb{C}$ of degree d. To see this we apply Theorem 5.6 (Limits of proper maps), which provides a subsequence with one of three possible types of behavior.

First, it might be the case that f_n converges to a constant. But this would imply that f_n has an attracting fixed point for all n sufficiently large, contrary to assumption.

Second, it might be the case that $f_n(z)$ converges to infinity for all but finitely many z. But for any $R > 0$ the Euclidean annulus

$$A(R) = \{z \ : \ 1 < |z| < R\}$$

is contained in V_n and encloses $P(f_n)$. By Theorem 2.1, $f_n^{-1}(A(R))$ contains a round annulus B_n enclosing $Q(f_n)$ with

$$\mathrm{mod}(B_n) = \frac{\mathrm{mod}(A(R))}{d} - O(1).$$

Since $\mathrm{diam}\, Q(f_n) = 1$, by choosing R sufficiently large we can assure the outer boundary of B_n is at distance at least 1 from 0. Thus $|f_n(z)| < R$ when $|z| < 1$, so this second possibility is also ruled out.

The remaining possibility is that f_n converges to a polynomial g of degree between 1 and d. But the critical points of f_n are contained in the ball $B(0, 1)$ for all n, so the limit g has degree exactly d.

Since f_n converges to the polynomial g, for all sufficiently large n there exist polynomial-like restrictions $f_n : U_n' \to V_n'$ of degree d with $\mathrm{mod}(U_n', V_n') \to \infty$ and $P(f_n) \subset B(0, 1) \subset U_n'$. The Julia sets of these polynomial-like maps are connected because the critical points do not escape from U_n'.

∎

Using the Koebe distortion theorem and allied results, the above theorem can be made quantitative (for example one can take $m_d(x) = (1 - 1/d)x - O(1)$.) We will only need the qualitative version above.

5.6   Univalent line fields

This section develops a particularly well-behaved class of holomorphic line fields, namely those which are univalent.

Definitions. A line field μ on a disk $V \subset \mathbb{C}$ is *univalent* if μ is the pullback of the horizontal line field in the plane under a univalent map $h : V \to \mathbb{C}$; that is, if $\mu = h^*(d\bar{z}/dz)$.

A holomorphic line field μ has a *zero* at z if $\phi(z) = 0$ where ϕ is a quadratic differential dual to μ near z. Since $h^*(d\bar{z}/dz)$ is dual to $h^*(dz^2)$ and $h' \neq 0$, *a univalent line field has no zeros.*

Here is a fairly general notion of an invariant line field. Let $f : U \to V$ be a nonconstant holomorphic map, and let μ be a line field on V. Then we say μ is *f-invariant* if $f^*\mu = \mu$ on $U \cap V$.

Theorem 5.13 *Let $f : U \to V$ admit a univalent invariant line field. Then f has no critical points in $U \cap V$.*

Proof. If $f'(z) = 0$ and $z \in U \cap V$, then by invariance μ has a zero at z.

∎

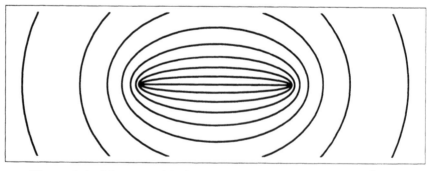

Figure 5.1. Meromorphic invariant line field for $f(z) = z^2 - 2$.

We will see that suitable expansions of a measurable line field about a point of almost continuity tend to a univalent line field in the limit. This theorem is a more precise application of the expansion philosophy, already used in the proof of Theorem 3.17 to obtain a meromorphic line field.

For our applications, the more precise version is needed because a dynamical system which arises as a limit of renormalization *might* admit an invariant meromorphic line field. For example, the polynomial $f(z) = z^2 - 2$ (with Julia set equal to the interval $[-2, 2]$) leaves invariant the line field dual to the meromorphic quadratic differential $\phi = dz^2/(z^2 - 4)$. See Figure 5.1.

However, this line field is not univalent near $J(f)$. Indeed, the preceding result shows that no restriction $f : U \to V$ of f to a neighborhood of its Julia set $J(f)$ can admit a univalent invariant line field, since the critical point $z = 0$ lies in $J(f)$.

This incompatibility between critical points and univalent line fields is a central theme in our approach to rigidity.

Definition. Let $(V_n, v_n) \to (V, v)$ in the space of disks $\mathcal{D}$, and let μ_n and μ be measurable line fields defined on V_n and V. (The support of each may be smaller.) We say μ_n *converges to* μ *in measure* if for all compact $K \subset V$ and all $\epsilon > 0$,

$$\text{area}(\{z \in K \ : \ |\mu(z) - \mu_n(z)| > \epsilon\}) \to 0$$

as $n \to \infty$.

Theorem 5.14 *Let* $f_n : (U_n, u_n) \to (V_n, v_n)$ *be a sequence of holomorphic maps between disks, converging to a non-constant map* $f :$ $(U, u) \to (V, v)$ *in the Carathéodory topology. Suppose* μ_n *is a sequence of* f_n*-invariant line fields on* V_n *converging in measure to* μ *on* V. *Then* μ *is* f*-invariant.*

Proof. Let $z \in U \cap V$ be any point such that $f'(z) \neq 0$ and μ is almost continuous at z and $f(z)$. Since almost every point in $U \cap V$ satisfies these conditions, it suffices to verify f-invariance at z.

By almost continuity, there is a small ball B centered at z so μ is nearly constant on most of B and on most of $f(B)$, and f' is nearly constant on B. By convergence in measure, when n is large, μ_n is nearly equal to $\mu(z)$ on most of B and to $\mu(f(z))$ on most of $f(B)$. By f_n-invariance, $\mu_n|f(B)$ is close to $\mu_n|B$ rotated by $f_n'(z)$.

Since $f'_n(z) \to f'(z)$, $\mu(f(z))$ is equal to $\mu(z)$ rotated by $f'(z)$, and therefore μ is f-invariant.

∎

Theorem 5.15 *If μ_n is a univalent line field on $(V_n, v_n) \to (V, v)$, then there is a subsequence such that μ_n converges in measure to a univalent line field μ on V.*

Proof. Write $\mu_n = h_n^*(d\bar{z}/dz)$ where h_n is univalent. Since the horizontal line field is invariant under translations and real dilations, we can arrange that $h_n(v_n) = 0$ and $|h'_n(v_n)| = 1$. By the Koebe principle, after passing to a subsequence, h_n converges to a univalent map $h : V \to \mathbb{C}$, so $\mu_n \to \mu = h^*(d\bar{z}/dz)$ in measure.

∎

Theorem 5.16 (Univalent promotion) *Let μ be a measurable line field on $\mathbb{C}$, and let x be a point of almost continuity of μ with $|\mu(x)| = 1$. Suppose $(V_n, v_n) \to (V, v)$ is a convergent sequence of disks, and $h_n : V_n \to \mathbb{C}$ is a sequence of univalent maps with $h'_n(v_n) \to 0$ and*

$$\sup \frac{|x - h_n(v_n)|}{|h'_n(v_n)|} < \infty.$$

Then there exists a subsequence such that $h_n^*(\mu)$ converges in measure to a univalent line field on V.

Remark. If $h_n(v_n) = x$, we need only require that $h'_n(v_n) \to 0$. In general we do not even require that the image of h_n contains x. Rather, the sup condition above guarantees that h_n carries V_n close enough to x that the line field μ is nearly parallel on most of the image.

Proof. After a preliminary rotation of the plane, we may assume that $\mu(x) = d\bar{z}/dz$. Let $\nu_n = h_n^*(d\bar{z}/dz)$; this line field is univalent, so after passing to a subsequence ν_n converges to a univalent line field ν on V. By assumption the Euclidean distance

$$d(x, h_n(v_n)) < \lambda |h'_n(v_n)|$$

for a constant λ independent of n.

We claim $h_n^*(\mu)$ converges to ν in measure. It suffices to show convergence on any closed ball $B \subset V$.

There is a connected open set V' containing v and B such that $V' \subset V_n$ for all n sufficiently large. The univalent maps $h_n|V'$ form a precompact family when suitably normalized, by the Koebe theorem. Thus for all n sufficiently large,

$$
\begin{aligned}
d(h_n(v_n), h_n(B)) &< C(B)|h_n'(v_n)|, \\
\operatorname{diam}(h_n(B)) &< C(B)|h_n'(v_n)| \text{ and} \\
c(B)|h_n'(v_n)|^2 &< \operatorname{area}(h_n(B)),
\end{aligned}
$$

for constants $C(B)$ and $c(B) > 0$ depending on B but independent of n. It follows that we may choose $r_n \to 0$ such that $h_n(B)$ is contained in a ball of radius r_n about x, and the area of $h_n(B)$ is greater than αr_n^2, for a constant α independent of n. (More precisely, we may take $r_n = (C(B) + \lambda)|h_n'(v_n)| \to 0$, and $\alpha = c(B)/(C(B) + \lambda)^2$.)

Since x is a point of almost continuity, the density of points in $h_n(B)$ where μ deviates from the horizontal line field by more than ϵ tends to zero as $n \to \infty$. By Koebe again, the density of points in B where $|h_n^*(\mu) - \nu_n| > \epsilon$ tends to zero as well. Thus $h_n^*(\mu)$ and ν_n converge in measure to the same limit ν.

■

Chapter 6

Polynomials and external rays

In this chapter we discuss polynomials and the combinatorial topology of the Julia set. This material is in preparation for §7, where we will use renormalization to break the Julia set of a quadratic polynomial into many connected pieces. These pieces can potentially touch at periodic cycles, so here we study the way in which the Julia set is connected at its periodic points.

6.1 Accessibility

Definitions. Let K be a *full nondegenerate continuum* in the complex plane. This means K is a compact connected set of cardinality greater than one and $\mathbb{C} - K$ is connected.

By the Riemann mapping theorem, there is a unique conformal isomorphism

$$\phi : (\mathbb{C} - \overline{\Delta}) \rightarrow (\mathbb{C} - K)$$

such that $\phi(z)/z \rightarrow \lambda > 0$ as $z \rightarrow \infty$.

For each angle $t \in \mathbb{R}/\mathbb{Z}$, the *external ray* $R_t \subset \mathbb{C}$ is defined by

$$R_t = \{\phi(r \exp(2\pi i t) \; : \; 1 < r < \infty\}.$$

An external ray R_t *lands* at a point $x \in \partial K$ if

$$\lim_{r \to 1+} \phi(r \exp(2\pi i t)) = x.$$

We call x a *landing point* and t a *landing angle*; t is an *external angle* for x. Traditionally x is called the *radial limit* of ϕ at $\exp(2\pi it)$.

Theorem 6.1 *The set of landing angles has full measure in $\mathbb{R}/\mathbb{Z}$.*

Proof. Let $A = \{z \ : \ 1 < |z| < 2\}$. By the Cauchy-Schwarz inequality,

$$\left(\int_A |\phi'(z)||dz|^2 \right)^2 \leq \left(\int_A 1|dz|^2 \right) \left(\int_A |\phi'(z)|^2|dz|^2 \right)$$
$$= \text{area}(A) \cdot \text{area}(\phi(A)) < \infty.$$

Therefore $\int_1^2 |\phi'(r\exp(2\pi it))|dr$ is finite for almost every t. It follows that the tail of R_t has finite length, and hence converges, for almost every t.

■

The following result is classical (see, e.g. [Car2, §313], [Garn, Cor 4.2]):

Theorem 6.2 (F. and M. Riesz) *For any set $E \subset \mathbb{R}/\mathbb{Z}$ of positive measure, there are two landing angles in E with different landing points.*

In other words, the radial limits of ϕ are nonconstant on any set of positive measure.

Definition. A point $x \in \partial K$ is *accessible* if there is a path γ in $\mathbb{C} - K$ converging to x.

Lindelöf's theorem shows a point is accessible if and only if it is accessible by a hyperbolic geodesic. Thus geodesics always follow a reasonably efficient route to the boundary, and do not become sidetracked in blind alleys.

Theorem 6.3 (Lindelöf) *Suppose $\phi(z) \rightarrow x$ as $z \rightarrow \exp(2\pi it)$ along a path δ in $\mathbb{C} - \overline{\Delta}$. Then the ray R_t also lands at x.*

See e.g. [Ah2, Theorem 3-5].
Combining these results, we have:

Corollary 6.4 *A point in ∂K is accessible if and only if it is the landing point of some ray.*

More precisely, if δ is a path in $\mathbb{C} - K$ converging to $x \in \partial K$, then $\gamma = \phi^{-1} \circ \delta$ converges to a point $\exp(2\pi i t) \in S^1$ and the ray R_t lands at x.

Proof. Clearly a landing point is accessible.

Now suppose $x \in \partial K$ is accessible, and $\delta : [0, 1) \to \mathbb{C} - K$ is path such that $\delta(s) \to x$ as $s \to 1$. Consider the lifted path $\gamma(s) = \phi^{-1} \circ \delta(s)$. Then γ accumulates on some connected subset C of the circle, and therefore R_t converges to x for almost every point $\exp(2\pi i t) \in C$. By the Theorem of F. and M. Riesz, C must reduce to a single point, say $C = \{\exp(2\pi i t)\}$ (this also follows from the Schwarz reflection principle).

Therefore $\gamma(s) \to \exp(2\pi i t)$ and $\phi(\gamma(s)) \to x$. By Lindelöf's theorem, the ray R_t also converges to x.

∎

Corollary 6.5 *The set of landing points is dense in K.*

Proof. It is easy to see the set of accessible points is dense, by considering for each x in $\mathbb{C} - K$ the nearest point to x in K.

∎

Theorem 6.6 *Suppose x is a point in ∂K such that $K - \{x\}$ has at least $n > 1$ connected components. Then at least n external rays land at x.*

Corollary 6.7 (Rays count components) *If n external rays land at x, where $1 \leq n < \infty$, then $K - \{x\}$ has n components.*

Proof. The n rays separate $K - \{x\}$ into n pieces, which is the most possible by the preceding Theorem.

∎

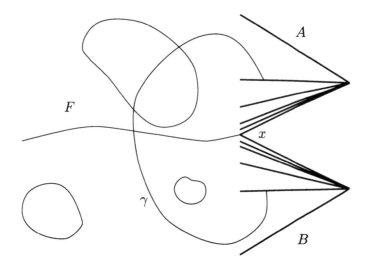

Figure 6.1. Accessibility of a cut point.

Proof of the Theorem. First suppose $n = 2$. Then we can write $K - \{x\} = A \cup B$, where A and B are disjoint sets closed in $\mathbb{C} - \{x\}$. By the Tietze extension theorem, there is a continuous function $\alpha : \mathbb{C} - \{x\} \to [0, 1]$ such that $\alpha^{-1}(0) = A$ and $\alpha^{-1}(1) = B$ (see [Roy]). A routine differential topology argument shows we can assume α is actually smooth. Let $F = \alpha^{-1}(r)$ for a regular value r in $(0, 1)$; then F is an embedded 1-dimensional submanifold of $\widehat{\mathbb{C}} - \{x\}$.

Let γ_1 be an arc whose interior lies in $\mathbb{C} - K$ and whose endpoints lie in A and B. Then γ_1 cuts $\widehat{\mathbb{C}} - K$ into two components U and V, where U is bounded in $\mathbb{C}$. If all components of $F \cap \overline{U}$ are compact, then we can join A to B by a path avoiding F, which contradicts the intermediate value theorem. Thus there is a submanifold $F_1 \subset F \cap \overline{U}$ which is diffeomorphic to $[0, \infty)$. Since F_1 is embedded in $\mathbb{C} - \{x\}$ and $\overline{U}$ is bounded, we must have $\overline{F_1} - F_1 = \{x\}$. Then F_1 provides a path tending to x in $\mathbb{C} - K$. (See Figure 6.1.)

This shows x is accessible and so at least one external ray, R_{t_1}, lands at x. To show two rays land, repeat the argument using an arc γ_2 joining A to B through $\mathbb{C} - K$ without crossing R_{t_1}. (Such an arc exists because R_{t_1} does not disconnect the plane). We obtain a path F_2 crossing γ_2 and tending to x. Let R_{t_2} be the corresponding ray landing at x as guaranteed by Lindelöf's theorem. Then R_{t_2} crosses γ_2, so these two rays are distinct.

The case of general $n > 1$ is similar. If $k < n$ rays land at x, then these rays divide K into k pieces $K_1, \ldots, K_k$, and $K_i - \{x\}$ must be disconnected for some i. We may then repeat the argument, writing $K_i - \{x\}$ as the union of disjoint closed sets A and B, and using an arc γ_{k+1} which passes through none of the k rays located so far. This produces a new path F_{k+1} to x, and hence a new ray, until we have found n of them.

∎

Remarks. The behavior of external rays is part of the general theory of prime ends; see [CL]. The theorems of F. and M. Riesz and of Lindelöf hold for more general classes of functions than those which arise as Riemann mappings.

6.2 Polynomials

Let $f : \mathbb{C} \to \mathbb{C}$ be a monic polynomial of degree $d > 1$. Recall the filled Julia set $K(f)$ is the set of all z for which $f^n(z)$ remains bounded as $n \to \infty$ (§3.1).

In this section we assume that $K(f)$ is connected (equivalently, $J(f)$ is connected).

To understand the combinatorics of the Julia set, it is often useful to imagine that $J(f)$ is a topological quotient of the unit circle S^1, in such a way that the dynamics of $z \mapsto z^d$ goes over to the dynamics of f. This image is not quite correct in general, because $J(f)$ need not be locally connected. Nevertheless many consequences of this heuristic are true.

Since the Julia set is connected, $K(f)$ is a full, nondegenerate continuum. As in the preceding section, we consider the Riemann mapping

$$\phi : (\mathbb{C} - \overline{\Delta}) \to (\mathbb{C} - K(f))$$

normalized so $\phi(z)/z \to \lambda > 0$ as $z \to \infty$. In fact, since f is monic, $\lambda = 1$, and this map is a conjugacy between z^d and f; that is,

$$\phi(z^d) = f(\phi(z)).$$

For any external ray R_t, its image $f(R_t) = R_{dt}$ is again an external ray. An external ray is *periodic* if $f^n(R_t) = R_t$, or equivalently $d^n t = t$, for some $n > 0$. The least such n is the *period* of R_t.

The following result is assembled from contributions of Douady, Hubbard, Sullivan and Yoccoz; see [DH1], [Dou1, §6], [Mil2, §18] and [Hub].

Theorem 6.8 *Every periodic external ray lands on a repelling or parabolic point for f. Conversely, let x be a repelling or parabolic periodic point for f. Then x is a landing point, and every ray landing at x is periodic with the same period.*

By Corollary 6.7 we have:

Corollary 6.9 *If x is a repelling or parabolic periodic point, then $K(f) - \{x\}$ has a finite number of components, equal to the number of rays landing at x.*

Remark. We do not know if an external ray can land at an irrationally indifferent periodic point x in the Julia set (a *Cremer point*). If it does, then $K(f) - \{x\}$ has infinitely many components.

Quadratic polynomials. Now suppose $f(z) = z^2 + c$ is a quadratic polynomial with connected Julia set. By tradition (see [DH1]), the landing point of R_0 is a repelling or parabolic fixed point of f called β, or the *zero angle fixed point*. The other fixed point of f is called α. When $f(z) = z^2 + 1/4$ (the only case with a multiple fixed point), we set $\alpha = \beta = 1/2$.

Theorem 6.10 *The β fixed point of a quadratic polynomial does not disconnect the filled Julia set.*

Proof. By Theorem 6.8, any ray R_t landing at β has the same period as R_0, namely one. But zero is the only fixed point of $t \mapsto 2t$ on $\mathbb{Z}/\mathbb{R}$, so only the zero ray lands at β. Then $K(f) - \beta$ is connected by Theorem 6.6.

∎

Corollary 6.11 *A repelling or parabolic fixed point* x *disconnects* $K(f)$ *if and only if* $x = \alpha \neq \beta$.

Proof. If x disconnects then $x = \alpha \neq \beta$ by the preceding result. Conversely, if α is parabolic or repelling, then at least two rays land at there by Theorem 6.8, and these separate $K(f) - \alpha$ into at least two pieces.

■

6.3 Eventual surjectivity

Consider the map $F : S^1 \to S^1$ given by $F(z) = z^d$, $d > 1$. It is easy to see that F is *locally eventually onto*: that is, for any nonempty open $U \subset S^1$, there is an $n > 0$ such that $F^n(U) = S^1$.

In this section we formulate a similar result for the Julia set of a polynomial.

Definition. Let $K \subset \mathbb{C}$ be a full nondegenerate continuum. A *cross cut* γ for K is the closure of an open arc in $\mathbb{C} - K$ which converges to ∂K at either end. Thus γ is either a closed arc joining two points of ∂K, or a topological circle meeting ∂K in a single point.

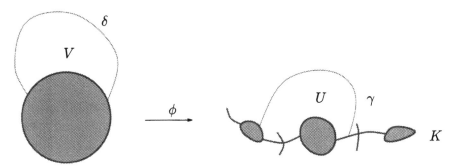

Figure 6.2. A crosscut for the filled Julia set.

Theorem 6.12 *Let* f *be a polynomial with connected filled Julia set* $K(f)$. *Let* $\gamma \subset \mathbb{C}$ *be a closed arc which is a cross cut for* $K(f)$, *and let* U *be the bounded component of* $\mathbb{C} - (K(f) \cup \gamma)$.

Then there exists an integer $n > 0$ such that the Julia set of f is contained in the bounded component of $\mathbb{C} - f^n(U)$.

Proof. Let $\phi : (\mathbb{C} - \overline{\Delta}) \to (\mathbb{C} - K(f))$ be the Riemann mapping normalized as before.

Suppose γ joins z_1 and z_2, two distinct points in ∂K. Let $\delta = \phi^{-1}(\gamma)$ and let $V = \phi^{-1}(U)$. By Corollary 6.4, the ends of δ converge to points t_1 and t_2 on S^1, and the corresponding external rays land at z_1 and z_2; thus $t_1 \neq t_2$. Consequently $\overline{V}$ contains an open interval on S^1. (See Figure 6.2.)

Therefore $S^1 \subset \overline{F^n(V)}$ for some $n > 0$, where $F(z) = z^d$. Since $F^n(V)$ is open, it contains a simple closed curve separating S^1 from infinity, so $f^n(U)$ disconnects the Julia set from infinity.

∎

Here is an application that will be used repeatedly in our study of the combinatorics of renormalization (§7).

Theorem 6.13 (Connectedness principle) *Let $f : \mathbb{C} \to \mathbb{C}$ be a polynomial with connected filled Julia set $K(f)$. Let $f^n : U \to V$ be a polynomial-like map of degree $d > 1$ with connected filled Julia set K_n. Then:*

1. *The Julia set of $f^n : U \to V$ is contained in the Julia set of f.*

2. *For any closed connected set $L \subset K(f)$, $L \cap K_n$ is also connected.*

Proof. The first claim is immediate from the fact that repelling periodic points of f^n are dense in the Julia set of $f^n : U \to V$. This follows from the fact that f^n is hybrid equivalent to a polynomial (Theorem 5.7).

For the second, suppose $L \cap K_n$ is not connected. Then there is a bounded component W of $\mathbb{C} - (L \cup K_n)$ such that $L \cap \partial W$ is a proper subset of ∂W (see Figure 6.3). Therefore we can construct an arc γ in $\overline{W}$ forming a cross cut for K_n. This cross cut can be chosen to lie arbitrarily close to K_n.

By Theorem 5.7, f^n is topologically conjugate to a polynomial of degree d near K_n. Using Theorem 6.12, we conclude that the region

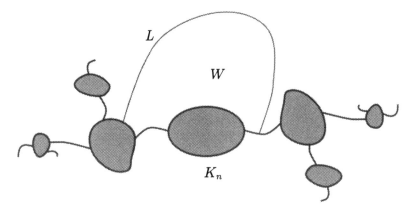

Figure 6.3. Proof of connectedness.

U between γ and K_n eventually maps onto an open set separating K_n from infinity. Since $U \subset W \subset K(f)$ and the filled Julia set $K(f)$ is full, we conclude that K_n lies in the interior of $K(f)$. But this is impossible, since $\partial K_n \subset \partial K(f)$. Thus $L \cap K_n$ is connected.

■

6.4 Laminations

A point in the Julia set of a polynomial can be the landing point of more than one external ray. Knowledge of the coincidence of external rays goes a long way towards determining the combinatorics of the polynomial. These coincidences are conveniently encoded by a lamination.

In this section we define the rational lamination of a polynomial with connected Julia set, and give a topological criterion for rational rays to land at the same point. This criterion will be used in the Appendix on quotient maps.

The theory of laminations is implicit in work of Douady and Hubbard and explicitly developed in [Th2]. Some authors use a different but closely related definition of lamination, emphasizing chords of the circle rather than equivalence classes.

Let $S^1 = \mathbb{R}/\mathbb{Z}$, and identify S^1 with the boundary of the unit disk Δ via the map $t \mapsto \exp(2\pi i t)$.

Definitions. A *lamination* $\lambda \subset S^1 \times S^1$ is an equivalence relation on circle such that the convex hulls of distinct equivalence classes are disjoint. One may form the convex hull of an equivalence class using either the Euclidean or hyperbolic metric on Δ; the results are homeomorphic. Indeed, the former corresponds to the Klein model for the hyperbolic plane.

The *support* of a lamination is the union of its nontrivial equivalence classes (those classes consisting of more than one point). A lamination is *finite* if its support is a finite set.

Let f be a monic polynomial with connected Julia set. The *rational lamination* $\lambda_{\mathbb{Q}}(f)$ is defined by $t \sim t'$ if $t = t'$, or if t and t' are rational and the external rays R_t and $R_{t'}$ land at the same point in the Julia set $J(f)$. It is easy to verify that $\lambda_{\mathbb{Q}}(f)$ *is* a lamination, using the fact that two simple closed curves on the sphere cannot cross at just one point.

Now let λ be a finite lamination with support Θ. We will give a condition which implies $\lambda \subset \lambda_{\mathbb{Q}}(f)$.

Let

$$\phi : (\mathbb{C} - \overline{\Delta}) \to (\mathbb{C} - K(f))$$

denote the Riemann mapping with $\phi(z)/z \to 1$ as in §6.2. For $t \in \mathbb{R}/\mathbb{Z}$, let $S_t = [1, \infty) \exp(2\pi i t)$.

A λ-*ray system* is a continuous map

$$\sigma : \bigcup_{t \in \Theta} S_t \to (\mathbb{C} - P(f))$$

such that:

1. there is an R such that $\sigma(z) = \phi(z)$ when $|z| > R$, and

2. $\sigma(z) = \sigma(z')$ if and only if $z = z'$ or $|z| = |z'| = 1$ and the corresponding angles t and t' are equivalent under λ.

Thus σ gives an embedding of the quotient of $\bigcup_\Theta S_t$ by the equivalence relation λ determines on the endpoints of the S_t's.

Two λ-ray systems σ_0 and σ_1 are *homotopic* if there is a continuous family of λ-ray systems σ_s, $s \in [0, 1]$ connecting them, and an R independent of s with $\sigma_s(z) = \phi(z)$ for $|z| > R$.

A λ-ray system σ_0 is *invariant* if there is a λ-ray system σ_1 homotopic to σ_0 such that

$$\sigma_0(z^d) = f(\sigma_1(z)).$$

The map σ_1 can be viewed as a lifting of $\sigma_0(z^d)$. Since $f : (\mathbb{C} - f^{-1}P(f)) \to (\mathbb{C} - P(f))$ is a covering map, invariance depends only on the homotopy class of σ_0.

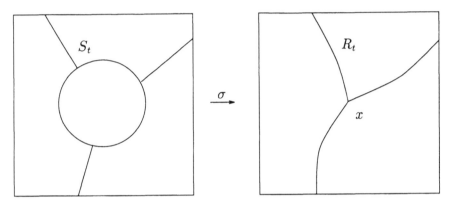

Figure 6.4. An invariant ray system.

Example. Let x be a fixed point of f lying outside the postcritical set $P(f)$, and suppose R_t lands at x for all $t \in \Theta \subset \mathbb{R}/\mathbb{Z}$, $|\Theta| > 1$. Let λ be the equivalence relation with the single nontrivial equivalence class Θ. Define σ on $\bigcup_\Theta S_t$ by $\sigma(z) = \phi(z)$ for $|z| > 1$ and $\sigma(z) = x$ when $|z| = 1$. Then σ is an f-invariant λ-ray system; see Figure 6.4.

Theorem 6.14 *If f admits an invariant λ-ray system, then λ is a subset of the rational lamination of f.*

We will need:

Theorem 6.15 *Let f be a polynomial of degree $d > 1$. Then exactly one of the following holds:*

1. *$f(z)$ is conjugate to z^d; or*

2. *$J(f)$ is connected and $J(f) \subset P(f)$; or*

3. $f^{-1}(P(f))$ meets the unbounded component Ω of $\mathbb{C} - P(f)$.

Proof. The map f is conjugate to z^d if and only if $|P(f)| \leq 2$. Setting aside this case, we may assume that Ω is a hyperbolic Riemann surface. If $f^{-1}(P(f))$ does not meet Ω, no finite critical point is attracted to infinity and therefore $K(f)$ is connected. Moreover $f^{-1}(\Omega) = \Omega$ so Ω lies outside the Julia set. Therefore $J(f) = \partial\Omega \subset P(f)$.

■

Proof of Theorem 6.14. Let Θ be the support of λ. Let σ_0 be an invariant λ-ray system, with a lift σ_1 homotopic to σ_0. Since $\sigma_0(z) = \sigma_1(z)$ for $|z|$ sufficiently large, the map $t \mapsto dt \bmod 1$ sends Θ into itself. Therefore every $t \in \Theta$ is preperiodic, hence rational, so R_t lands at a preperiodic point $x(t) \in J(f)$ by Theorem 6.8. Our goal is to show that $x(t) = x(t')$ whenever $(t, t') \in \lambda$.

Let $\Theta' \subset \Theta$ denote those angles which are periodic under $t \mapsto dt$. Note that an invariant λ-ray system for f is also an invariant λ-ray system for f^n, and $\lambda_\mathbb{Q}(f^n) = \lambda_\mathbb{Q}(f)$. Thus we may assume every angle in Θ' is fixed by $t \mapsto dt$, and $d\Theta = \Theta'$, since these conditions may be achieved by replacing f with f^n for an appropriate value of n.

Let Ω denote the unbounded component of $\mathbb{C} - P(f)$, and let $\Omega' = f^{-1}(\Omega)$.

Claim 1: The region Ω' is a proper subset of Ω.

Otherwise by Theorem 6.15 $f : \Omega \to \Omega$ is a covering map and $\partial\Omega$ is equal to $J(f)$ or to a single point. Suppose for example $\partial\Omega = J(f)$; then the map $h : \mathbb{H} \to \Omega$ given by $h(z) = \phi(\exp(-2\pi iz))$ presents the upper halfplane as the universal cover of Ω, and we may lift each σ_i to a map

$$\tilde{\sigma}_i : \bigcup_\Theta S_t \to \mathbb{H}$$

such that $h \circ \tilde{\sigma}_i = \sigma_i$. There is a lift of f to a map

$$F : \mathbb{H} \to \mathbb{H}$$

of the form $F(z) = dz + k$ for some integer k, for which

$$\tilde{\sigma}_0(z^d) = F(\tilde{\sigma}_1(z)).$$

But for $|z|$ sufficiently large the mappings $\tilde{\sigma}_i$ coincide and send each S_t to a vertical line in the upper halfplane. Since only one vertical line is invariant under F while $|\Theta| > 1$, we obtain a contradiction. The argument when $f(z)$ is conjugate to z^d is similar. This establishes Claim 1.

Now by invariance we may construct a sequence of λ-ray systems σ_n such that $\sigma_n(z^d) = f(\sigma_{n+1}(z))$. Assume $\sigma_0(z) = \phi(z)$ for $|z| > R > 1$; then $\sigma_n(z) = \phi(z)$ when $|z| > R_n = R^{1/d^n} \to 1$ as $n \to \infty$. In other words, for n large σ_n carries most of S_t onto most of the external ray R_t.

For $t \in \Theta$, let

$$F_{n,t} = S_t \cap \{z \ : \ |z| \le R_n\}.$$

Let $E_{n,t}$ denote the Euclidean diameter of $\sigma_n(F_{n,t})$.

Claim 2: As $n \to \infty$, $E_{n,t} \to 0$.

To establish this claim, first assume $t \in \Theta'$, so $dt = t$. Let $\rho(z)|dz|$ denote the hyperbolic metric on Ω, and let

$$H_{n,t} = \int_{F_{n,t}} \rho(\sigma_n(z))|\sigma_n'(z)||dz|$$

denote the parameterized hyperbolic length of the image of $F_{n,t}$. Since $f : \Omega' \to \Omega$ is a covering map, and σ_{n+1} is a lift of σ_n, the length of image of $F_{n+1,t}$ relative to the hyperbolic metric on Ω' is equal to $H_{n,t}$. Since Ω' is a proper subset of Ω and inclusions are contracting, we have $H_{n+1,t} \le H_{n,t}$. Moreover the contraction of the inclusion $\Omega' \to \Omega$ is uniform on any compact subset of Ω', so either $H_{n,t} \to 0$ or $\sigma_n(F_{n,t})$ eventually leaves every compact subset of Ω. In the former case the Euclidean diameter $E_{n,t} \to 0$ because the hyperbolic length bounds the Euclidean length. In the latter case $\sigma_n(F_{n,t})$ tends to the boundary of Ω, so its Euclidean diameter tends to zero because the ratio of the hyperbolic to Euclidean metric tends to infinity (cf. Theorem 2.3).

This establishes Claim 2 when $t \in \Theta'$. Now for any $t \in \Theta$,

$$f(\sigma_{n+1}(F_{n+1,t})) = \sigma_n(F_{n,dt}),$$

and $dt \in \Theta'$, so the Euclidean diameter of the image of $F_{n,t}$ tends to zero in this case as well (because $F_{n+t,t}$ is connected).

Finally we show Claim 2 implies the theorem. Indeed,

$$\sigma_n(R_n \exp(2\pi it)) = \phi(R_n \exp(2\pi it)) \to x(t)$$

as $n \to \infty$. On the other hand, whenever $(t, t') \in \lambda$,

$$\sigma_n(F_{n,t}) \cup \sigma_n(F_{n,t'})$$

is a connected set of Euclidean diameter at most

$$E_{n,t} + E_{n,t'} \to 0$$

containing $\sigma_n(R_n \exp(2\pi it))$ to $\sigma_n(R_n \exp(2\pi it'))$. Thus $x(t) = x(t')$ as desired.

∎

Chapter 7

Renormalization

Renormalization is a tool for the study of nonlinear systems whose essential form is repeated at infinitely many scales.

For a quadratic polynomial $f(z) = z^2 + c$, this repetition of form takes place when a high iterate f^n sends a small neighborhood of the critical point $z = 0$ over itself by degree two. Then a suitable restriction of f^n is quadratic-like, and we can hope to reduce the analysis of f to that of its iterate f^n. This reduction works best when the quadratic-like map has connected Julia set (so the critical point does not escape). The passage from the quadratic map f to the quadratic-like map f^n is an instance of renormalization.

This chapter develops the combinatorics of renormalization for quadratic polynomials. We begin by showing that when it exists, a renormalization of f^n is essentially unique. Then we study the n small Julia sets associated to f^n, and how they fit together as n varies. Using this combinatorics we define *simple* renormalization, which will be our main focus in the sequel. The chapter concludes with some examples of renormalizable quadratic polynomials.

7.1 Quadratic polynomials

Definitions. Let $f(z) = z^2 + c$ be a quadratic polynomial with connected Julia set. A *quadratic-like map* is a polynomial-like map of degree two.

The map f^n is *renormalizable* if there are open disks U and V in

$\mathbb{C}$ such that the critical point $0 \in U$ and

$$f^n : U \to V$$

is a quadratic-like map with connected Julia set. (Equivalently, $f^{nk}(0) \in U$ for all $k \geq 0$.) [1]

The choice of a pair (U, V) as above is a *renormalization* of f^n. Let

$$\mathcal{R}(f) \quad = \quad \{n \geq 1 \ : \ f^n \text{ is renormalizable}\}.$$

The integers n which appear in $\mathcal{R}(f)$ are the *levels* of renormalization.

Theorem 7.1 (Uniqueness of renormalization) *Any two renormalizations of f^n have the same filled Julia set.*

Proof. Let $f^n : U^1 \to V^1$ and $f^n : U^2 \to V^2$ be two renormalizations of f^n, with filled Julia sets K^1 and K^2. By Theorem 6.13, $L = K^1 \cap K^2$ is connected, and clearly $f^n(L) = L$. Let U be the component of $U^1 \cap U^2$ containing L, and let $V = f^n(U)$. By Theorem 5.11, $f^n : U \to V$ is polynomial-like with filled Julia set equal to L, and of degree two because the critical point $z = 0$ lies in L. Since the degrees of all three maps are the same, we have $L = K^1 = K^2$.

■

Next we collect together notation that will be used in the sequel. Suppose for each n in $\mathcal{R}(f)$ we have chosen a renormalization

$$f^n : U_n \to V_n.$$

Then:

- P_n, J_n and K_n denote the postcritical set, Julia set and filled Julia set of the quadratic-like map $f^n : U_n \to V_n$. By assumption J_n and K_n are connected, so $P_n \subset K_n$.

---

[1]Milnor has suggested the following notation. Let f be a polynomial with a distinguished critical point ω. Then f is *n-renormalizable about ω* if there are open disks U and V containing ω such that $f^n : U \to V$ is polynomial-like with connected Julia set, and ω is the only critical point of f in U. Thus our terminology "f^n is renormalizable" is shorthand for "f^n is 1-renormalizable about $\omega = 0$."

- $K_n(i) = f^i(K_n)$ for $i = 1, \dots, n$. These *small filled Julia sets* are cyclically permuted by f. Note that $K_n(n) = K_n$.

- $P_n(i) = K_n(i) \cap P(f)$ is the *ith small postcritical set*. We have $P(f) = \{\infty\} \cup \bigcup_{i=1}^{n} P_n(i)$.

- $J_n(i) = \partial K_n(i)$ is the *ith small Julia set*.

- $K_n = K_n(1) \cup \dots \cup K_n(n)$ is the union of the small filled Julia sets at level n. We have $f(K_n) = K_n$.

- $J_n = J_n(1) \cup \dots \cup J_n(n)$.

- $V_n(i) = f^i(U_n)$ for $i = 1, \dots, n$. Then the quadratic-like map f^n is factored as

$$U_n \xrightarrow{f} V_n(1) \xrightarrow{f} \dots \xrightarrow{f} V_n(n) = V_n,$$

where the first map $U_n \rightarrow V_n(1)$ is proper of degree two and the remaining maps are univalent.

- $U_n(i)$ is the component of $f^{i-n}(U_n)$ contained in $V_n(i)$. We will see that $f^n : U(i) \rightarrow V(i)$ is quadratic-like (Theorem 7.2).

- $P_n'(i)$, $J_n'(i)$ and $K_n'(i)$ are defined by $P_n'(i) = -P_n(i)$, and so on. Each primed object has the same image under f as its unprimed companion, and for $i \neq n$ the primed and unprimed objects are disjoint.

By Theorem 7.1, the filled Julia set K_n of a renormalization is *canonical*, even though the choice of U_n and V_n may not be. As a consequence, $K_n(i)$, $J_n(i)$, $P_n(i)$ and their primed companions are also canonical.

Next we investigate the deployment of the small filled Julia sets $K_n(1), \dots K_n(n)$.

Theorem 7.2 *Let f^n be renormalizable. Then for $i = 1, \dots, n$,*

$$f^n : U_n(i) \rightarrow V_n(i)$$

is quadratic-like with filled Julia set $K_n(i)$. Similarly,

$$(-f^n) : U_n'(i) \rightarrow V_n'(i)$$

is quadratic-like with filled Julia set $K'_n(i)$. Both maps are holomorphically conjugate to $f^n : U_n \to V_n$.

Proof. The map $f^{n-i} : V_n(i) \to V_n(n)$ is univalent, and it conjugates $f^n : U_n(i) \to V_n(i)$ to $f^n : U_n \to V_n$, which is quadratic-like. Therefore $f^n : U_n(i) \to V_n(i)$ is quadratic-like, and its filled Julia set is $K_n(i)$ because $f^{n-i}(K_n(i)) = K_n$. Similarly, $(-f^{n-i}) : U'_n(i) \to U_n$ conjugates f^n to $(-f^n)$ because $f(-z) = f(z)$. ∎

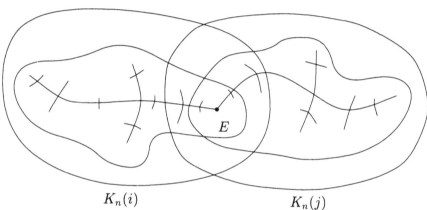

$K_n(i)$ $\qquad\qquad\qquad\qquad\qquad$ $K_n(j)$

Figure 7.1. The small Julia sets are almost disjoint.

Theorem 7.3 (Almost disjoint Julia sets) *Suppose two distinct small filled Julia sets $K_n(i)$ and $K_n(j)$ have a nonempty intersection. Then $K_n(i) \cap K_n(j) = \{x\}$, where x is a repelling fixed point of f^n.*

Proof. Let $E = K_n(i) \cap K_n(j)$. Then $f^n(E) \subset E$ and E is connected by Theorem 6.13.

Let W be the component of $U_n(i) \cap U_n(j)$ containing E (see Figure 7.1), and let $W' = f^n(W)$. By Theorem 5.11, $f^n : W \to W'$ is polynomial-like, and of degree one because $K_n(i) \neq K_n(j)$. By the Schwarz lemma, E consists of a single repelling fixed point for f^n. ∎

Theorem 7.4 *Suppose f^n is renormalizable. Then any attracting or indifferent periodic point and any periodic component of the interior of $K(f)$ is contained in $K_n(i)$ for a unique i. Its period is divisible by n.*

Corollary 7.5 *Every periodic point of f with period less than n is repelling.*

Proof. Let x be an attracting periodic point, or an indifferent periodic point lying in the Julia set. Then x lies in $P(f)$ by basic facts in rational dynamics (Corollary 3.7). Therefore $x \in K_n(i)$ for some i. The small filled Julia sets meet only at repelling points, if at all (Theorem 7.3), so this i is unique. By uniqueness, if $f^p(x) = x$ then $f^p(K_n(i)) = K_n(i)$, and therefore the period p is a multiple of n.

If D is a component of the interior of $K(f)$ of period p, then D is an attracting or parabolic basin, or a Siegel disk. In the attracting or parabolic case, D is contained in the unique $K_n(i)$ containing the corresponding attracting or parabolic periodic point. If D is a Siegel disk, then $\partial D \subset P(f)$, and thus $P_n(i) \cap \partial D$ is open and nonempty in ∂D for some i. Since $P_n(i)$ is invariant under f^{np} and $f^{np}|D$ is an irrational rotation, it is easy to see that $\partial D \subset P_n(i)$. Then $D \subset K_n(i)$ because $K_n(i)$ is full, and this i is unique because the interiors of the small filled Julia sets are disjoint. Uniqueness again implies $n|p$.

An indifferent periodic point in the interior of $K(f)$ is the center of a unique Siegel disk, so this case is also covered.

∎

Theorem 7.6 (Least common renormalization) *If f^a and f^b are renormalizable, then so is f^c, where c is the least common multiple of a and b. The corresponding filled Julia sets satisfy $K_c = K_a \cap K_b$.*

Proof. Define U_a^* by

$$U_a^* = \{z \in U_a : f^{aj}(z) \in U_a \text{ for } j = 1,\ldots c/a - 1\};$$

then $f^c : U_a^* \to V_a$ is polynomial-like of degree $2^{c/a}$. Define U_b^* in the same way, with b in place of a.

By Theorem 6.13, the set $L = K_a \cap K_b$ is connected. Let U_c be the component of $U_a^* \cap U_b^*$ containing L, and let $V_c = f^c(U_c)$. By Theorem 5.11, the map $f^c : U_c \to V_c$ is polynomial-like, with filled Julia set L.

The critical point $z = 0$ lies in $f^i(L) = f^i(K_a) \cap f^i(K_b)$ if and only if $a|i$ and $b|i$; that is, i must be a multiple of c. Therefore f^c has a single critical point in L, and since L is connected, $f^c : U_c \to V_c$ is polynomial-like of degree 2. Therefore $c \in \mathcal{R}(f)$ and $K_c = L = K_a \cap K_b$.

■

Corollary 7.7 *If f^a and f^b are renormalizable, and a divides b, then $K_a \supset K_b$.*

7.2   Small Julia sets meeting at periodic points

The combinatorics of renormalization is simplest when the small Julia sets are actually disjoint. We have seen, however, that the small Julia sets can touch at repelling periodic points. In this section we will show that the periods of these touching points tend to infinity as the level of renormalization tends to infinity.

Theorem 7.8 (High periods) *Given a period p, there are only finitely many n in $\mathcal{R}(f)$ such that the small filled Julia set K_n contains a periodic point of f of period p.*

Theorem 7.8 is often a good substitute for disjointness. The main point in the proof is to show that for $n \in \mathcal{R}(f)$ sufficiently large, K_n does not contain either fixed point of f.

To begin an analysis of fixed points, let $f(z) = z^2 + c$ be a quadratic polynomial with connected Julia set such that both fixed points α and β of f are repelling.

Theorem 7.9 *The external rays landing at α are permuted transitively by f. The external rays landing at $-\alpha$ separate β from the critical point of f.*

Proof. By Theorem 6.8, a finite number q of external rays land at the α fixed point of f. If a ray R_t lands at α, then so does $f(R_t) = R_{2t}$; since f is locally injective at α, it is clear that f permutes the rays landing there. We will show this permutation is transitive.

The rays landing at α are forward invariant, and they divide the complex plane into q open components $P_1, \ldots, P_q$; we may assume P_q contains the critical point and $-\alpha$. The preimages of these rays land at α and $-\alpha$, dividing the plane into $2q - 1$ pieces $Q_1, Q_2, \ldots Q_{2q-1}$. Since only the piece P_q is subdivided by the rays landing at $-\alpha$, we may label these new pieces so $Q_i = P_i$ for $1 \leq i < q$ and $Q_q \subset P_q$ contains the critical point. (See Figure 7.2.)

Assume $i < q$, so $P_i = Q_i$ does not contain the critical point. Then f maps P_i univalently to another piece $f(P_i) = P_j$. The angle between the rays bounding $f(P_i)$ is twice the angle between the rays bounding P_i. Since the angle cannot increase without bound, eventually P_i maps onto P_q. Therefore the rays landing at α are permuted transitively.

Since β is fixed by f while $f(P_i)$ is disjoint from P_i when $i < q$, it follows that β is contained in P_q. Similarly β is not contained in Q_q, since $f(Q_q) = P_i$ for some $i < q$ (containing the critical value). Thus β is contained in Q_j for some $j > q$, and therefore the rays landing at $-\alpha$ separate β from the critical point in Q_q.

∎

Remark. The preceding argument contains the beginnings of the Yoccoz puzzle, discussed more fully in §8.2.

We now consider $f(z) = z^2 + c$ such that f^n is renormalizable for some $n > 1$. Then the Julia set of f is connected, and by Corollary 7.5 f has two repelling fixed points α and β. (In other words, the c's for which a proper iterate of $z^2 + c$ is renormalizable lie outside the main cardioid of the Mandelbrot set.)

Theorem 7.10 *If f^n is renormalizable and $n > 1$, then the small filled Julia set K_n does not contain the β fixed point of f.*

Proof. Suppose to the contrary that K_n contains β. Then $f(K_n) = K_n(1)$ also contains β, since $f(\beta) = \beta$. By Theorem 7.3, the small

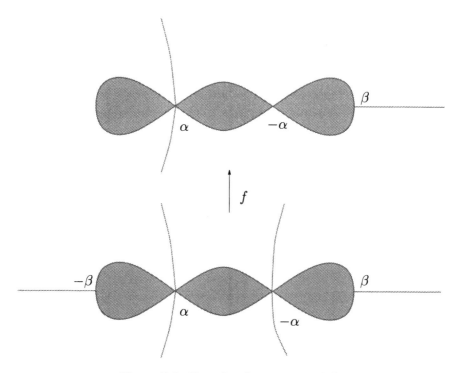

Figure 7.2. Rays landing at α and β.

filled Julia sets meet in at most one point, so K_n does not contain α. Consequently K_n does not contain $-\alpha$ either. Thus K_n is disjoint from the rays landing at $-\alpha$. But these rays separate the critical point from β, which is impossible because K_n is connected and contains both these points.

∎

Theorem 7.11 *Suppose f^n is renormalizable for $n > 1$, and $\alpha \in K_n$. Let ρ be the number of components of $K_n - \{\alpha\}$. Then*

$$n\rho \leq q,$$

where q is the number of external rays landing at α.

Combining the last two results, we obtain:

Corollary 7.12 *For $n \in \mathcal{R}(f)$, $n > q$, the small filled Julia set K_n does not contain any fixed point of f.*

The idea of the proof of the Theorem 7.11 is straightforward: the set $\mathcal{K}_n$ is separated into $n\rho$ components by α, and these components are locally cyclically permuted by f. So starting with a single external ray landing at α, we obtain $n\rho$ such rays by applying f to it. We will use Riemann mappings and Lindelöf's theorem to formalize this argument.

Proof of Theorem 7.11. Let

$$\Theta = \{t_1, \ldots t_q\} \subset \mathbb{R}/\mathbb{Z}$$

denote the external angles of the rays landing at the α fixed point of f. By Theorem 7.9, f permutes these rays transitively; in other words, the map $t \mapsto 2t \bmod 1$ gives a cyclic permutation of Θ. Let $F : \Theta \to \Theta$ denote the *inverse* of this permutation.

The map F can be described geometrically as follows: starting with a ray R_t landing at α, form the set $f^{-1}(R_t)$; this consists of one ray R_s landing at α, and another $R_{s+1/2}$ landing at $-\alpha$. Then $F(t) = s$.

Recall $\mathcal{K}_n = \bigcup_{i=1}^{n} K_n(i)$ is the union of the small Julia sets at level n. Let $\mathcal{L}_n = f^{-1}(\mathcal{K}_n)$.

Both $\mathcal{K}_n$ and $\mathcal{L}_n$ are full continua. Indeed, $\mathcal{K}_n$ is a fan of n copies of K_n joined at α, and

$$\mathcal{L}_n = \mathcal{K}_n \cup -K_n(1) \cup \ldots \cup -K_n(n-1)$$

is obtained from $\mathcal{K}_n$ by attaching $n-1$ copies of K_n to it at $-\alpha$.

For appropriate Riemann mappings $\pi_{\mathcal{L}}$ and $\pi_{\mathcal{K}}$ we obtain a commutative diagram of conformal coverings maps:

$$
\begin{array}{ccc}
\mathbb{C} - \overline{\Delta} & \xrightarrow{\ z^2\ } & \mathbb{C} - \overline{\Delta} \\
{\scriptstyle \pi_{\mathcal{L}}}\downarrow & & \downarrow{\scriptstyle \pi_{\mathcal{K}}} \\
\mathbb{C} - \mathcal{L}_n & \xrightarrow{\ f\ } & \mathbb{C} - \mathcal{K}_n.
\end{array}
$$

We will denote external rays for the Riemann mapping $\pi_{\mathcal{K}}$ by R'_t; thus

$$R'_t = \pi_{\mathcal{K}}((1,\infty)\exp(2\pi i t)).$$

Since α separates K_n into ρ components, it also separates each $K_n(i)$ into ρ pieces; thus $\mathcal{K}_n - \{\alpha\}$ has exactly $n\rho$ components. By Theorem 6.6, there are exactly $n\rho$ external rays of the form R'_t landing at α under the Riemann mapping $\pi_{\mathcal{K}}$; denote their external angles by $\Theta' \subset \mathbb{R}/\mathbb{Z}$.

Define a map $F' : \Theta' \to \Theta'$ as follows. Given an external ray R'_t landing at α under the Riemann mapping $\pi_{\mathcal{K}}$, consider its inverse image $f^{-1}(R'_t)$. This set consists of two paths γ and $-\gamma$ landing at α and $-\alpha$ respectively. The path γ lies outside $\mathcal{L}_n$ and hence outside $\mathcal{K}_n$. By Lindelöf's theorem (that is, Corollary 6.4), $\pi_{\mathcal{K}}^{-1}(\gamma)$ converges to a point $z = \exp(2\pi i s)$ on the unit circle and the external ray R'_s lands at α. Set $F'(t) = s$.

Finally, define $h : \Theta \to \Theta'$ by a similar construction: given an external ray R_t in the complement of the filled Julia set $K(f)$, let $h(t) = t'$ where $\pi_{\mathcal{K}}^{-1}(R_t)$ terminates at $\exp(2\pi i t')$.

The conclusion of the argument amounts to verifying:

(1) h gives a semiconjugacy between F and F'; that is, the diagram

$$
\begin{array}{ccc}
\Theta & \xrightarrow{\ F\ } & \Theta \\
{\scriptstyle h}\downarrow & & \downarrow{\scriptstyle h} \\
\Theta' & \xrightarrow{\ F'\ } & \Theta'
\end{array}
$$

is commutative; and

(2) F' is a cyclic permutation of Θ'.

Indeed, (1) and (2) imply h is surjective, so $|\Theta| \geq |\Theta'|$ which says simply that $q \geq n\rho$.

Proof of (1). Let $t \in \Theta$. Then R_t and $R'_{h(t)}$ both land at α. In addition, the preimages of R_t and $R'_{h(t)}$ under $\pi_{\mathcal{K}}$ land at the same point on the unit circle — namely $\exp(2\pi i h(t))$.

The external ray $R_{F(t)}$ is the unique component of the preimage of R_t under f which terminates at α. Similarly, there is a unique component γ of the preimage of $R'_{h(t)}$ which also terminates at α.

Since $f \circ \pi_{\mathcal{L}}(z) = \pi_{\mathcal{K}}(z^2)$, we can construct $R_{F(t)}$ by first lifting R_t via $\pi_{\mathcal{K}}$, then choosing the appropriate component of its preimage under z^2, and then projecting by $\pi_{\mathcal{L}}$. Note that the other component of the preimage projects by $\pi_{\mathcal{L}}$ to a ray landing at $-\alpha$.

The same considerations apply to the construction of γ from $R'_{h(t)}$. Since γ and $R_{F(t)}$ both terminate at α (rather than $-\alpha$), and their lifts by $\pi_{\mathcal{K}}$ land at the same point on S^1, the correct further preimages under z^2 also land at the same point on S^1. Equivalently, $\pi_{\mathcal{L}}^{-1}(\gamma)$ and $\pi_{\mathcal{L}}^{-1}(R_{F(t)})$ land at the same point $w \in S^1$. See Figure 7.3, which depicts an example in which $n = \rho = 2$.

Now let $i : (\mathbb{C} - \mathcal{L}_n) \to (\mathbb{C} - \mathcal{K}_n)$ denote the inclusion mapping, and let $\tilde{i}$ denote its lift to the uniformizations of the domain and range by $\pi_{\mathcal{L}}$ and $\pi_{\mathcal{K}}$ respectively. The image of $\tilde{i}$ is $\mathbb{C} - (\overline{\Delta} \cup E)$, where $E = \pi_{\mathcal{K}}^{-1}(\mathcal{L}_n)$.

Since $\mathcal{L}_n - \mathcal{K}_n$ is cut off from $\mathcal{K}_n$ by finitely many rays landing at $-\alpha$, while γ and $R_{F(t)}$ land at α, we have $\tilde{i}(w) \notin \overline{E}$. By Schwarz reflection, $\tilde{i}$ extends continuously to a neighborhood of w. Thus the images of γ and $R_{F(t)}$ under $\tilde{i} \circ \pi_{\mathcal{L}}^{-1}$ land at the same point on the circle.

Equivalently, the preimages of $R_{F(t)}$ and of γ under $\pi_{\mathcal{K}}$ terminate at the same place. By definition, the form preimage terminates at $h(F(t))$, while the latter terminates at $F'(h(t))$. Thus $F'(h(t)) = h(F(t))$.

Proof of (2). The rays R'_t for $t \in \Theta'$ divide the set $\mathcal{K}_n - \{\alpha\}$ into its $n\rho$ components, $P_1, \ldots, P_{n\rho}$. We may assume the critical point lies in P_1. Arguing as in Theorem 7.9, one may show that for each

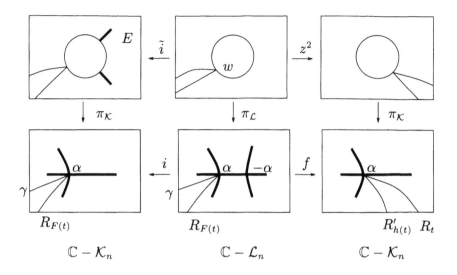

Figure 7.3. Factoring Riemann mappings.

$i > 1$, $f(P_i) = P_j$ for some j, and if $f(P_i)$ lies in the sector bounded by rays R'_s and R'_t, then P_i lies in the sector bounded by rays $R'_{F(s)}$ and $R'_{F(t)}$. This reduces the proof to checking that each P_i eventually maps onto P_1.

We know $P_1 \subset K_n$ and each $P_i \subset K_n(j)$ for some j, $1 \le j \le n$. Thus $f^{n-j}(P_i) \subset K_n$, so it suffices to verify the claim when $P_i \subset K_n$. Consider the polynomial g which is hybrid equivalent to the polynomial-like map $f^n : U_n \to V_n$. If α corresponds to the β fixed point of g, then $K_n - \{\alpha\}$ has only one component so we are done. Otherwise the α fixed point of f corresponds to the α fixed point of g. Then the desired result follows from the assertion that g cyclically permutes the rays landing at *its* α fixed point (Theorem 7.9).

                                                      ■

Proof of Theorem 7.8 (High periods). Let w be a point of period p for f. We will show $w \in K_n$ for only finitely many $n \in \mathcal{R}(f)$.

Suppose there is an a in $\mathcal{R}(f)$ with $a > p$ and $w \in K_a$; then $f^p(w) = w \in K_a(p)$. By Theorem 7.3, $\{w\} = K_a(p) \cap K_a$ and w is a repelling fixed point of f^a. Let g be the quadratic polynomial to which $f^a : U_a \to V_a$ is hybrid equivalent. Then w corresponds to either the α or β fixed point of g.

First suppose w corresponds to β. We claim $w \notin K_b$ for all $b > a$ in $\mathcal{R}(f)$. Indeed, Theorem 7.6 guarantees that f^c is renormalizable and $K_c = K_b \cap K_a$ where $c = \mathrm{lcm}(a, b)$. The renormalization of f^c provides a renormalization of $g^{c/a}$ whose filled Julia set does not contain the β fixed point of g, by Theorem 7.10. It follows that K_c, and therefore K_b, fails to contain w.

Now suppose w corresponds to the α fixed point of g. Let q be the number of external rays for g landing at its α fixed pont. By Corollary 7.12, α is disjoint from the filled Julia set of any renormalization of g of level greater that q. So the same reasoning shows K_b fails to contain w for any $b > qa$.

Thus the set of $n \in \mathcal{R}(f)$ such that $w \in K_n$ is finite. Since there are only finitely many periodic points of period p, the theorem follows.

∎

7.3 Simple renormalization

In this section we introduce some terminology and numerical invariants for renormalization. In particular we define *simple renormalization*, which plays a fundamental role in the sequel.

Definitions. Let $f(z) = z^2 + c$ be a quadratic polynomial, and let n belong to $\mathcal{R}(f)$. For $1 \leq i \leq n$, $f^n : U_n(i) \to V_n(i)$ is a polynomial-like map, hybrid equivalent to a quadratic polynomial g with connected Julia set. The fixed points of g may be labeled α and β in accordance with §6.2. Let $\alpha_n(i)$ and $\beta_n(i)$ denote the corresponding fixed points in $K_n(i)$.

By Theorem 7.3 if two small Julia sets meet, say $K_n(i) \cap K_n(j) = \{p\}$ for $i \neq j$, then $p = \alpha_n(i)$ or $p = \beta_n(i)$ (but not both – since p is a repelling fixed point of f^n, $\alpha_n(i) \neq \beta_n(i)$).

Theorem 7.13 (Same type) *All intersections of small Julia sets at a given level n occur at the same type of fixed point (α or β).*

A more precise statement of Theorem 7.13 is the following: it is never the case that $K_n(i) \cap K_n(j) = \{\alpha_n(i)\}$ while $K_n(i') \cap K_n(j') = \{\beta_n(i')\}$.

Proof. First note that $f(\alpha_n(i)) = \alpha_n(i+1)$ and $f(\beta_n(i)) = \beta_n(i+1)$ (where $n+1$ is interpreted as 1). For $i < n$ this is immediate from the fact the f gives a conformal conjugacy between $f^n : U_n(i) \to V_n(i)$ and $f^n : U_n(i+1) \to V_n(i+1)$. For $i = n$ it follows from the fact that f^n fixes $\alpha_n(n)$ and $\beta_n(n)$.

Now suppose two small Julia sets meet at $\alpha_n(j)$ while two others meet at $\beta_n(k)$; we will find a contradiction. Since f cyclically permutes the collections $K_n(i)$, $\alpha_n(i)$ and $\beta_n(i)$, every point of the form $\alpha_n(i)$ or $\beta_n(i)$ belongs to at least two small Julia sets. Thus $|F| \le n$ where

$$F = \bigcup_{i=1}^{n} \{\alpha_n(i), \beta_n(i)\}.$$

Form a graph with one vertex for each point of F and edges $e(i)$ joining $\alpha_n(i)$ to $\beta_n(i)$ for $i = 1, \dots, n$. This graph has no more vertices than edges, so it includes at least one cycle $< e(i_1), \dots, e(i_k) >$. Since $\alpha_n(i_1) \ne \beta_n(i_1)$, we have $k > 1$. Then $L = K_n(i_2) \cup \dots \cup K_n(i_k)$ is connected since adjacent sets in the union meet. By Theorem 6.13, $L \cap K_n(i_1)$ is also connected. But $L \cap K_n(i_1) = \{\alpha_n(i_1), \beta_n(i_1)\}$, a contradiction.

■

Types of renormalization. A renormalization of f^n is of:

> α-*type*, if some pair of small Julia sets meet at their α fixed points;
>
> β-*type*, if some pair meet at their β fixed points; and of
>
> *disjoint type*, if the small Julia sets are disjoint.

By Theorem 7.13, every renormalization is of exactly one of these types.

A renormalization is *simple* if it is of β-type or disjoint type. Equivalently, whenever two small Julia sets meet, they do so at their β fixed points.

A renormalization is *crossed* if it is not simple. Crossed is synonymous with α-type. The terminology is meant to suggest that the small Julia sets cross at their α fixed points.

Let

$$\mathcal{SR}(f) \;=\; \{n \in \mathcal{R}(f) \;:\; \text{the renormalization of } f^n \text{ is simple}\}.$$

Theorem 7.14 *If $a \in \mathcal{R}(f)$ and $b \in \mathcal{SR}(f)$, then a divides b or b divides a.*

Proof. Let d be the greatest common divisor of a and b. If $d = \min(a,b)$ then $a|b$ or $b|a$ and we are done.

Otherwise, d is less than both a and b, so $K_a \neq K_a(d)$ and $K_b \neq K_b(d)$. Note that K_a meets K_b because they both contain the critical point $z = 0$. Therefore $f^i(K_a)$ meets $f^i(K_b)$ for any $i > 0$. In particular, $K_a(d)$ meets $K_b(d)$.

The sequences of sets $f^i(K_a)$ and $f^i(K_b)$ are periodic with periods a and b respectively. Some multiple of b is congruent to $d \bmod a$, and vice-versa, so K_b meets $K_a(d)$ and K_a meets $K_b(d)$.

Thus $L = K_b \cup K_a(d) \cup K_b(d)$ is connected. By Theorem 6.13, $L \cap K_a$ is also connected. Since $K_a(d) \cap K_a$ is at most a single point, we conclude $(K_b \cup K_b(d)) \cap K_a$ is connected. As both K_b and $K_b(d)$ meet K_a, connectedness implies $K_b \cap K_b(d) \cap K_a$ is nonempty. But $b \in \mathcal{SR}(f)$, so K_b and $K_b(d)$ meet at their β fixed points. Thus the β fixed point of $f^b : U_b \to V_b$ lies in $K_a \cap K_b$.

Let c be the least common multiple of a and b. By Theorem 7.6, f^c is renormalizable, and $K_c = K_a \cap K_b$. But then the polynomial g to which $f^b : U_b \to V_b$ is hybrid equivalent admits a proper renormalization whose Julia set contains the β fixed point of g. This contradicts Theorem 7.10.

∎

Corollary 7.15 *The set $\mathcal{SR}(f)$ is totally ordered with respect to division.*

By Corollary 7.7, we have:

Corollary 7.16 *The sets K_n form a nested decreasing sequence as n increases through values in $\mathcal{SR}(f)$. Consequently, for any pair $a < b$ in $\mathcal{SR}(f)$, each small Julia set at level b is contained in some small Julia at level a, and the same is true for the small postcritical sets.*

The following result will be useful for constructing simple renormalizations in §8.

Theorem 7.17 *Let f^a be simply renormalizable, and let g be the quadratic polynomial to which $f^a : U_a \to V_a$ is hybrid equivalent. Suppose g^b is simply renormalizable. Then f^c is simply renormalizable, for $c = ab$.*

Proof. To say g is hybrid equivalent to f^a means there is a quasi-conformal map $\psi : W_f \to W_g$ conjugating f^a to g, where W_f and W_g are neighborhoods of K_a and $K(g)$.

Suppose $g^b : U_b \to V_b$ provides a simple renormalization of g. Replacing U_b and V_b by their preimages under a higher iterate of g^b if necessary, we can assume $U_b \subset V_b \subset W_g$. Then $(U_c, V_c) = (\psi^{-1}(U_b), \psi^{-1}(V_b))$ provides a renormalization of f^c, where $c = ab$.

We must verify that the renormalization of f^c is simple; we may assume $b > 1$. By Corollary 7.7, $K_c \subset K_a$ (this is also clear from the construction.) Thus every $K_c(i)$ is contained in some $K_a(i')$.

Now suppose $K_c(i) \cap K_c(j) = \{x\}$. We have $K_c(i) \subset K_a(i')$ and $K_c(j) \subset K_a(j')$ for some i' and j'. If $i' \neq j'$, then x must be the β fixed point of $K_a(i')$ because the renormalization of f^a was simple. But then $K_c(i)$ also contains β, contrary to Theorem 7.10. Therefore $i' = j'$, and $f^a : K_a(i') \to K_a(i')$ is topologically conjugate to $g : K(g) \to K(g)$. It follows that $K_c(i) - K_c(j)$ is connected because the renormalization of g^b is simple. Therefore the renormalization of f^c is also simple.

■

Multiplicity and ramification. If f^n is renormalizable, its *multiplicity* μ_n is the number of small filled Julia sets meeting K_n, including K_n itself. The multiplicity is one if and only if the renormalization is of disjoint type. It is the same as the number of small filled Julia sets in any component of $\mathcal{K}_n$.

The *ramification* ρ_n is the number of components of the set $K_n - \bigcup_{i \neq n} K_n(i)$. This is the same as the number of components of $K_n(j) - \bigcup_{i \neq j} K_n(i)$ for any j.

The ramification is one if and only if the renormalization is simple. In fact, $K_n \cap \bigcup_{i \neq n} K_n(i)$ is equal to $\{\alpha_n(n)\}$, $\{\beta_n(n)\}$ or $\emptyset$ according to whether the renormalization is of α-type, β-type of disjoint type. For a crossed renormalization, $\alpha_n(n)$ is repelling and $\rho_n > 1$ is the number of components of $K_n - \alpha_n(n)$ (which is finite by Corollary 6.9).

7.4 Examples

To illustrate these results, we present (without proofs) several examples of renormalizable quadratic polynomials.

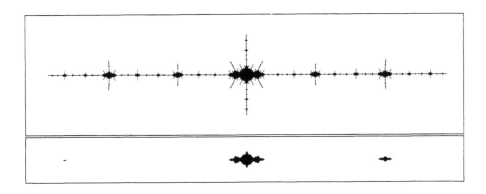

Figure 7.4. The small Julia sets are disjoint.

I. Let $f(z) = z^2 - 1.772892\ldots$ satisfy $f^6(0) = 0$. For this map, f^3 is renormalizable and its quadratic-like restriction $f^3 : U_3 \to V_3$ is hybrid equivalent to $z^2 - 1$. Its Julia set is depicted in Figure 7.4, with $K_3(1) \cup K_3(2) \cup K_3(3)$ drawn below. In this case the small Julia sets are disjoint.

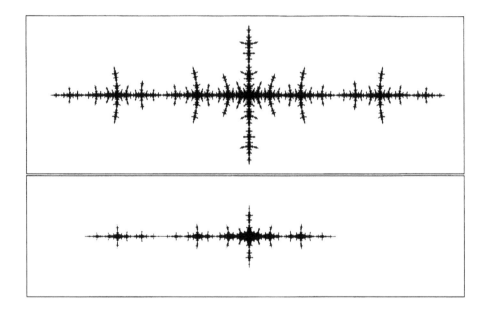

Figure 7.5. The Feigenbaum map: the small Julia sets touch.

II. Let $f(z) = z^2 - 1.401155\ldots$ be the Feigenbaum polynomial, illustrated in Figure 7.5; then f^2 is renormalizable, and again $K_2(1) \cap K_2(2) = \{x\}$ where x is a fixed point of f.

The Feigenbaum polynomial has many special properties with respect to renormalization; for example, f^{2^n} is renormalizable for any $n > 0$, and all such renormalizations are hybrid equivalent to f itself. This map can be described as the limit of the "cascade of period doublings" for $z^2 + c$ as c decreases along the real axis, starting at zero.

It turns out that the point x shared by $K_2(1)$ and $K_2(2)$ does *not* belong to $P(f)$; indeed the finite postcritical set is a Cantor set on which f acts injectively without periodic cycles. This absence of periodic cycles in $P(f)$ (other than infinity) is a general feature of infinitely renormalizable maps, as we will see below.

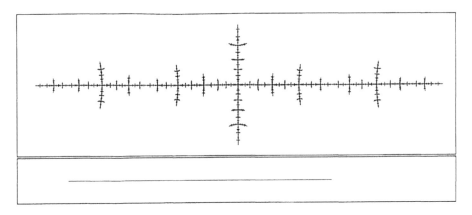

Figure 7.6. The small postcritical sets touch.

III. Let $f(z) = z^2 - 1.54368\ldots$ satisfy $f^3(0) = f^4(0)$; then f^2 is renormalizable and hybrid equivalent to $z^2 - 2$.

The Julia set of f appears in Figure 7.6; drawn below it is $K_2(1) \cup K_2(2)$, a pair of intervals meeting at $x = f^3(0)$, a fixed point of f which is also in the postcritical set. So not only do the small Julia sets meet in this example, the small postcritical sets $P_2(1)$ and $P_2(2)$ also meet.

IV. A perhaps surprising example is given by $f(z) = z^2 + 0.389007\ldots + 0.215851i \ldots$; for this map, the critical point has periodic six, and f^2, f^3 and f^6 are all renormalizable. This shows the set of n for which f^n is renormalizable does not have to be totally ordered with respect to division. (The set of *simple* renormalizations, however, is so ordered.)

The filled Julia set of f is rendered at the top of Figure 7.7. Let B be the closure of the immediate basin of attraction of $z = 0$; then the pictures at the bottom depict $\bigcup f^{-2n}(B)$ and $\bigcup f^{-3n}(B)$. The closure of the connected component in black containing zero is K_2 at the left, and K_3 at the right. The map f^2 is hybrid equivalent to the "rabbit" $z^2 - 0.122561\ldots + 0.744861\ldots$, while f^3 is hybrid equivalent to $z^2 - 1$.

V. Let $f(z) = z^2 + 0.419643\ldots + 0.606291i\ldots$; for this map, f^2 is

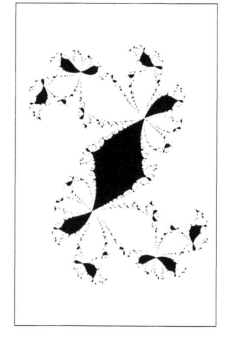

Figure 7.7. Here f^2, f^3 and f^6 are all renormalizable.

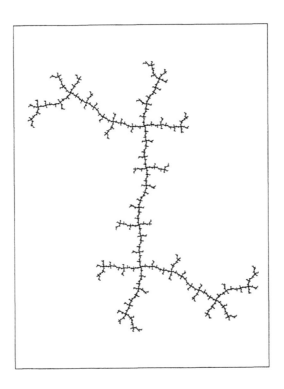

Figure 7.8. The small Julia set of f^2 is an quasiarc.

renormalizable and hybrid equivalent to $z^2 - 2$, but no higher iterate of f is renormalizable. The Julia set of f appears in Figure 7.8. The small Julia sets $J_2(1)$ and $J_2(2)$ are a pair of quasiarcs (quasiconformal images of intervals), running horizontally and vertically and crossing at the α fixed of f.

These example are classified as follows.

I: $\mathcal{R}(f) = \{1, 3, 6\}$. The renormalization f^3 is of disjoint type, while that of f^6 is of β-type.

II: $\mathcal{R}(f) = \{1, 2, 4, \ldots, 2^n, \ldots\}$. Each renormalization f^{2^n} is of β-type with multiplicity $\mu_{2^n} = 2$.

III: $\mathcal{R}(f) = \{1, 2\}$. The renormalization f^2 is of β-type with multiplicity two.

IV: $\mathcal{R}(f) = \{1, 2, 3, 6\}$. The renormalization f^2 is of α-type with multiplicity two and ramification three. Similarly f^3 is of α-type with $\mu_3 = 2$, $\rho_3 = 2$. Finally f^6 is of β-type with $\mu_6 = 6$.

V: $\mathcal{R}(f) = \{1, 2\}$; the renormalization f^2 is of α-type with $\mu_2 = \rho_2 = 2$.

The renormalizations of f^2 and f^3 in example IV, and of f^2 in example V, are crossed. The others are simple.

Example V shows that a quadratic polynomial need not admit any simple renormalizations (other than f itself), even when f^n is renormalizable for some $n > 1$.

Tuning. The "tuning" construction of Douady and Hubbard (unpublished) provides many more examples of renormalizable polynomials.

Quite informally, the idea of tuning is the following. Start with a polynomial $g(z) = z^2 + c$ such that the critical point $z = 0$ is periodic of period p. There is a unique Riemann mapping from the immediate basin of attraction U of $z = 0$ to the unit disk Δ which conjugates $g^p(z)$ to z^2. This mapping provides "internal angles" for points in the boundary of U.

Let $h(z) = z^2 + c'$ be a second polynomial with connected Julia set. Replace U with a copy of $K(h)$, identifying points in ∂U and points in $\partial K(h)$ whose internal and external angles correspond.

Carry out a similar surgery on each component of the interior of the filled Julia set $K(g)$. The result is a new polynomial $f(z) = z^2 + c''$, the "tuning" of g by h.

As long as h is not $z^2 + 1/4$, f^p is simply renormalizable and $f^p : U_p \to V_p$ is hybrid equivalent to h.

For more details, see [Dou1], [Dou2], [Mil1].

It would be interesting to have a parallel systematic construction of polynomials admitting crossed renormalizations.

Chapter 8

Puzzles and infinite renormalization

This chapter presents basic facts concerning infinitely renormalizable quadratic polynomials.

The set of parameters c for which z^2+c is infinitely renormalizable seems to be quite thin. Nevertheless, these mappings are of central interest, both for their internal symmetries and because they are yet to be well-understood. For example, these are the only quadratic polynomials for which the no invariant line fields conjecture is still open.

We begin with properties of such polynomials which follow from the material already developed. Then we describe the "Yoccoz puzzle", a Markov partition for the Julia set, and use it to show that an infinitely renormalizable map admits infinitely many *simple* renormalizations. Next we summarize fundamental results of Yoccoz, Lyubich and Shishikura, showing that a quadratic polynomial which is only *finitely* renormalizable carries no invariant line field on its Julia set. Finally we give a lamination criterion for renormalizability.

8.1 Infinite renormalization

Definition. A quadratic polynomial is *infinitely renormalizable* if f^n is renormalizable for infinitely many positive integers n.

121

Theorem 8.1 *Let f be infinitely renormalizable. Then:*

1. *All periodic cycles of f are repelling.*

2. *The filled Julia set of f has no interior.*

3. *The intersection $\bigcap_{\mathcal{R}(f)} \mathcal{J}_n$ contains no periodic points.*

4. *The finite postcritical set $P(f) \cap \mathbb{C}$ contains no periodic points.*

5. *For any $n \in \mathcal{R}(f)$, $P_n(i)$ and $J_n(j)$ are disjoint when $i \neq j$.*

Proof. Corollary 7.5 states that all cycles of period less than $n \in \mathcal{R}(f)$ are repelling. Since f is infinitely renormalizable, every periodic cycle is repelling. Thus $K(f) = J(f)$ because a nonempty open set in the filled Julia set entails an attracting, parabolic or indifferent cycle.

Let x be a periodic point of f. By Theorem 7.8, the forward orbit of x meets J_n for only finitely many $n \in \mathcal{R}(f)$. Thus x is disjoint from $\mathcal{J}_n$ for all n sufficiently large, and therefore $\bigcap \mathcal{J}_n$ contains no periodic points.

Since $P(f) \cap \mathbb{C} \subset \bigcap \mathcal{J}_n$, the postcritical set also contains no periodic points.

For $i \neq j$, $J_n(i)$ and $J_n(j)$ can only meet at a periodic point by Theorem 7.3. Therefore $P_n(i) \subset J_n(i)$ is disjoint from $J_n(j)$.

∎

Since the Julia set and filled Julia set are equal in the infinitely renormalizable case, we have $K(f) = J(f)$, $K_n(i) = J_n(i)$ and $\mathcal{K}_n = \mathcal{J}_n$ for all n in $\mathcal{R}(f)$. For simplicity we will use only the J notation when we consider infinitely renormalizable polynomials.

Theorem 8.2 (Small Julia sets attract) *Let f be infinitely renormalizable. Then for any n in $\mathcal{R}(f)$, and for almost every x in the Julia set of f, the forward orbit of x lands in $\mathcal{J}_n$.*

Remark. Since $\mathcal{J}_n = \bigcup_1^n J_n(i)$ is forward invariant, once an iterate of x lands there it remains in $\mathcal{J}_n$ for all future iterations.

Proof. The Julia set of f is not the whole sphere, so $d(f^k(x), P(f)) \to 0$ as $i \to \infty$ for almost every x in $J(f)$ (by Theorem 3.9). The post-critical set is partitioned into disjoint compact pieces $P_n(1), \ldots P_n(n) = P_n$ which are permuted by f. Therefore the forward orbit of x accumulates on P_n (as well as every other block of the partition).

For all k sufficiently large, when $f^k(x)$ is closer to P_n than to the rest of $P(f)$, then so is $f^{k+n}(x)$. Thus there is an iterate $f^k(x)$ such that $d(f^{k+nj}(x), P_n) \to 0$ as $j \to \infty$. Since P_n is a compact subset of U_n, for j large enough $y = f^{k+nj}(x)$ does not escape from U_n under iteration of f^n. Therefore $y \in J_n \subset \mathcal{J}_n$.

■

8.2 The Yoccoz jigsaw puzzle

In this section we describe a Markov partition for the dynamics of a quadratic polynomial, introduced by Yoccoz. We will use the language of tableau, developed earlier by Branner and Hubbard in their closely related work on cubic polynomials [BH]. For more details the reader is referred to [Mil3], [BH], [Hub] and [Yoc].

Definitions. Let $f(z) = z^2 + c$ be a quadratic polynomial with connected Julia set and *both fixed points repelling*. For simplicity, we also assume that *the forward orbit of the critical point is disjoint from the α fixed point of f*.

Let ϕ be the Riemann mapping from $\mathbb{C} - \bar{\Delta}$ to $\mathbb{C} - K(f)$, normalized so $\phi'(\infty) = 1$ as in §6.2. Consider the disk D bounded by the image of the circle $\{|z| = 2\}$ under ϕ. This disk encloses $K(f)$ and is cut into $q > 1$ pieces by the external rays landing at the α fixed point of f (compare §6).

Following [Mil3], we denote these pieces by $P_0(c_0), \ldots, P_0(c_{q-1})$, where $c_i = f^i(0)$ and c_i lies in $P_0(c_i)$. They form the "puzzle pieces" of depth zero. Each piece is a closed disk, whose boundary consists of the α fixed point, segments of external two rays and part of ∂D.

The puzzle pieces at depth $d + 1$ are defined inductively as the components of $f^{-1}(P)$, where P ranges over all puzzle pieces at depth d.

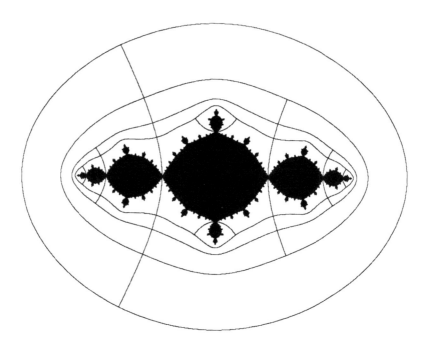

Figure 8.1. Puzzle pieces for $f(z) = z^2 - 1$.

The puzzle pieces at depth d have disjoint interiors and cover the Julia set. As the depth increases, the puzzles become successively finer: a piece at depth $d + 1$ is contained in a unique piece at depth d.

Example. The first few pieces in the puzzle for $f(z) = z^2 - 1$ are depicted in Figure 8.1.

Every point z in $K(f)$ which does not lie in the grand orbit of α is contained in a unique puzzle piece $P_d(z)$ at depth d. Clearly $P_d(f(z)) = f(P_{d+1}(z))$. The *tableau* for z is the array of pieces $P_d(f^k(z))$ for $d, k \geq 0$; it records the symbolic dynamics of z relative to the puzzle partitions. (Usually one adds more information to the tableau that will not be needed here.)

The renormalization period. The pieces $P_d(f^k(0)) = P_d(c_k)$ form the *critical tableau*; it is *periodic* if for some $n > 1$, $P_d(c_n) = P_d(0)$ for all depths d. The least such n is the *period of the critical tableau*. Since the pieces $P_0(0), \ldots P_0(c_{q-1})$ are always distinct, the period must be greater than one.

Similarly, we define the *renormalization period* of f to be the least $n > 1$ such that f^n is simply renormalizable.

By convention, if the tableau is aperiodic or if f has no proper simple renormalizations, we define the corresponding period to be ∞.

Theorem 8.3 *Let f be a quadratic polynomial with connected Julia set and both fixed points repelling, such that the forward orbit of the critical point avoids the α fixed point of f.*

Then the period of the critical tableau $P_d(c_k)$ is equal to the renormalization period of f.

Proof. Let $n > 1$ be the period of the critical tableau, assumed to be finite. We will first show that f^n is simply renormalizable, so n is greater than or equal to the renormalization period of f.

Since the period is exactly n, for d large enough the pieces $P_d(c_1)$, $\ldots$, $P_d(c_{n-1})$ have interiors disjoint from $P_d(0)$. Therefore $f^n : P_{d+n}(0) \to P_d(0)$ is a proper map of degree two.

Now by slightly thickening these pieces, one obtains a quadratic-like map with connected Julia set (see [Mil3, Lemma 2] for details).

We must check that the renormalization is simple. The small filled Julia set K_n is contained in $P_{d+n}(c_0)$, so $K_n(i)$ is contained in $P_{d+n-i}(c_i)$. By our choice of d these puzzle pieces have disjoint interiors for $i = 1, \ldots, n$. Two puzzle pieces whose interiors are disjoint can only meet at a point in the grand orbit of α. On the other hand, $K_n(i)$ and $K_n(j)$ can only meet at a repelling fixed point of f^n. Thus $K_n(i)$ and $K_n(j)$ can only touch at α. If they do, then since $f(\alpha) = \alpha$, all the small filled Julia sets meet at α and $n = q$, the number of external rays landing at α.

By Theorem 7.11, $n\rho_n \le q$, where ρ_n is the ramification of f^n (which in this case is equal to the number of components of $K_n - \{\alpha\}$). Since $n = q$, the ramification $\rho_n = 1$ and therefore f^n is simply renormalizable.

Now suppose f^n is simply renormalizable, for $n > 1$. To complete the proof, we will show that the period of the critical tableau is less than or equal to n.

First note that K_n is contained in $P_0(0)$. This is clear if K_n does not contain α. If K_n meets α, then so does $K_n(1)$, and thus α is the β fixed point of K_n (by our assumption of simplicity). Therefore $K_n - \{\alpha\}$ is connected, so it lies in single component of $K(f) - \alpha$, and consequently $K_n \subset P_0(0)$.

Next we claim $K_n \subset P_d(0)$ for all depths d. Indeed, K_n is connected, so for any $k > 0$ the set $f^{-kn}(K_n)$ has a unique component containing K_n. Thus K_n is contained in a unique puzzle piece at depth kn, which must coincide with $P_{kn}(0)$ because $0 \in K_n$. For any d we can choose k so $kn > d$, from which it follows that $K_n \subset P_d(0)$.

On the other hand, $c_n = f^n(0) \in K_n$ as well, so $P_d(c_n) = P_d(0)$ for all d. Therefore the critical tableau has period less than or equal to n.

∎

Remark. When the critical tableau has period n, one has $P_d(c_{i+n}) = P_d(c_i)$ for all i and d; see [Mil3, Lemma 2]).

8.3   Infinite simple renormalization

Next we give a combinatorial application of puzzles.

Theorem 8.4 *If f is infinitely renormalizable, then f admits an infinite sequence of simple renormalizations.*

The proof depends on:

Theorem 8.5 *Let $f(z) = z^2 + c$ be a quadratic polynomial, and let $E_1, \ldots, E_n$ be disjoint closed connected subsets of the filled Julia set $K(f)$. Suppose the critical point $z = 0$ lies in E_n, $f(E_n) \subset E_1$, and $f(E_i) \subset E_{i+1}$ for $i < n$.*

Then f^n is simply renormalizable, and $P_n(i) \subset E_i \subset K_n(i)$.

Proof. If $n = 1$ the result is immediate, so assume $n > 1$.

The sets E_i are disjoint and permuted by f, so their union contains no periodic point of period less than n. Since $P(f) \subset \bigcup E_i$, every periodic point of period less than n is repelling, by the same argument as in the proof of Theorem 7.4.

In particular, both fixed points of f are repelling, so we may construct the Yoccoz puzzle for f.

We claim that the critical tableau for f is periodic. To see this, first note that E_i is contained in the interior of the puzzle piece $P_0(c_i)$, since it contains $c_i = f^i(0)$ and does not meet α. Since E_i is connected and contained in $f^{-1}(E_{i+1})$, it follows by induction that $E_i \subset P_d(c_i)$ for every depth d. On the other hand, both 0 and c_n lie in E_n, so $P_d(0) = P_d(c_n)$ for all n and d. Thus the critical tableau is periodic, with period $a > 1$ dividing n.

By Theorem 8.3, f^a is simply renormalizable. If $a = n$, the remaining assertion of the Theorem are easily verified.

Otherwise, note that K_a consists of those points which remain in $P_d(0)$ under forward iteration of f^a, for some sufficiently large depth d. Therefore $E_a, E_{2a}, \ldots, E_n \subset K_a$. Now repeat the argument replacing f with the quadratic polynomial g to which f^a is hybrid equivalent near K_a. Applying Theorem 7.17 we eventually obtain a simple renormalization of f^n.

∎

Proof of Theorem 8.4. For each level of renormalization n in $\mathcal{R}(f)$, let κ_n denote the number of components of $\mathcal{K}_n$ and let μ_n denote the multiplicity of the renormalization (the maximal number of filled Julia sets meeting at a single point — see §7.3.) Then $n = \kappa_n \mu_n$.

We claim κ_n tends to infinity. Indeed, let x be a point where two or more small filled Julia sets at level n meet; then x is a repelling periodic point of f, and κ_n is at least as large as the period of x. By Theorem 7.8 (High periods), the period of x tends to infinity as n tends to infinity.

Next we show f^a is simply renormalizable for $a = \kappa_n$. To this end, let $E_1, \ldots, E_a$ denote the components of $\mathcal{K}_n$, ordered so that the critical point 0 lies in E_a and $f(E_i) = E_{i+1}$. Then the hypotheses of Theorem 8.5 are satisfied, so f^a is simply renormalizable. Since $\kappa_n \to \infty$, the map f admits an infinite sequence of simple renormalizations.

■

8.4   Measure and local connectivity

In this section we summarize results of Lyubich, Shishikura and Yoccoz which are proved using the Yoccoz puzzle.

Theorem 8.6 (Yoccoz) *Let $f(z) = z^2 + c$ be a quadratic polynomial such that*

1. *the Julia set $J(f)$ is connected,*

2. *f has no indifferent cycle, and*

3. *f is not infinitely renormalizable.*

Then $J(f)$ is locally connected.

If, in addition, f has no attracting cycle, then c lies in the boundary of the Mandelbrot set M and M is locally connected at c.

See [Yoc]; a detailed proof of local connectivity of $J(f)$ under slightly stronger assumptions can be found in [Mil3].

Corollary 8.7 *If the Julia set of $f(z) = z^2 + c$ carries an invariant line field, then f^n is simply renormalizable for infinitely many n.*

Proof. By Theorems 4.8 and 4.9, if f admits an invariant line field, then c lies in a non-hyperbolic component of the interior of the Mandelbrot set and every periodic cycle of f is repelling. So by Yoccoz's result, f is infinitely renormalizable, and by Theorem 8.4 infinitely many of these renormalizations are simple.

■

An alternative route to the Corollary above is given by the following result:

Theorem 8.8 (Lyubich, Shishikura) *If $f(z) = z^2 + c$ has no indifferent cycles and $J(f)$ has positive measure, then f is infinitely renormalizable.*

See [Lyu4].

Here is the skeleton of Lyubich's argument. Applying the Yoccoz puzzle, a new type of polynomial-like mapping, and a version of Theorem 2.16, Lyubich first proves:

Theorem 8.9 *Let f be a quadratic polynomial with both fixed points repelling. Then either the Julia set of f has measure zero, or f^n is renormalizable for some $n > 1$.*

Now suppose f has no indifferent cycles and $J(f)$ has positive measure. By the result above, f^a is renormalizable for some $a > 1$. To establish the theorem, one need only show that for any such a there is a $c > a$ such that f^c is also simply renormalizable.

By Theorem 5.7, there is a quadratic polynomial g to which f^a is hybrid equivalent. By a generalization of Theorem 8.2, almost every point in $J(f)$ eventually lands in J_a, so J_a has positive measure. Quasiconformal maps in the plane preserve sets of positive measure [LV, §IV.1.4], so $J(g)$ has positive measure. Since f has no indifferent cycle, both fixed points of g are repelling, and thus g^b is renormalizable for some $b > 1$. By Theorem 7.17, f^c is renormalizable for $c = ab$.

Remark. Actually, in the arguments of both Yoccoz and Lyubich, periodicity of the critical tableau is used to construct renormalizations. So these renormalizations are always simple by Theorem 8.3.

8.5   Laminations and tableaux

In this section we show the tableau of a quadratic polynomial is determined by the external angles of the inverse images of the α fixed point. It follows that renormalizability can be checked by looking at external rays; this fact will be used in Appendix B.

Recall from §6.4 that a lamination is an equivalence relation on the circle such that the convex hulls of distinct equivalence classes are disjoint. As in that section, we will identify the circle $S^1 = \mathbb{R}/\mathbb{Z}$ with the boundary of the disk via the map $t \mapsto \exp(2\pi i t)$.

Let $f(z) = z^2 + c$ be a quadratic polynomial with connected Julia set, whose fixed points are labeled α and β as in §6.2.

Definition. The α-*lamination* $\lambda_\alpha(f) \subset \lambda_\mathbb{Q}(f)$ is the subset of the rational lamination corresponding to rays which land in the inverse orbit of α. That is, $(t, t') \in \lambda_\alpha(f)$ if and only if $t = t'$ or t and t' are both rational, the external rays R_t and $R_{t'}$ land at the same point $z \in J(f)$, and $f^n(z) = \alpha$ for some $n \geq 0$.

Now suppose both fixed points α and β of f are repelling, and the forward orbit of the critical point of f is disjoint from α. Then the critical tableau $P_n(c_k)$ for the Yoccoz puzzle is well-defined.

We will show that the critical tableau can be reconstructed from the α-lamination of f. To make this precise, we will construct a model tableau $G_{n,k}$ canonically from $\lambda_\alpha(f)$.

Let $F : S^1 \to S^1$ be defined by $F(t) = 2t$. Since every point in the inverse image of α is prefixed, there is a natural stratification

$$\lambda_\alpha(f) = \bigcup_0^\infty \lambda_\alpha^d(f), \quad \lambda_\alpha^0(f) \subset \lambda_\alpha^1(f) \subset \lambda_\alpha^2(f) \subset \dots ,$$

given by

$$\lambda_\alpha^0(f) = \lambda_\alpha(f) \cap \{(t, t') \; : \; t = t' \text{ or } t \text{ and } t' \text{ are periodic under } F\},$$

and

$$\lambda_\alpha^d(f) = \lambda_\alpha(f) \cap \{(t, t') \; : \; (F^d(t), F^d(t')) \in \lambda_\alpha^0(f)\}$$

for $d > 0$. The lamination $\lambda_\alpha^0(f)$ has a unique nontrivial equivalence class, corresponding to the finite set of rays landing at α. Similarly $\lambda_\alpha^d(f)$ corresponds to the rays landing in $f^{-d}(\alpha)$.

For each lamination $\lambda_\alpha^d(f)$, let $\Lambda_d \subset \overline{\Delta}$ be the union of the convex hulls of equivalence classes in $\lambda_\alpha^d(f)$. A *gap at depth* d is the closure of a component of $\overline{\Delta} - \Lambda_d$. Let $\mathcal{G}_d$ be the set of such gaps.

For any gap G at depth $d + 1$, there is a unique map G' at level d such that $F(G \cap S^1) = G' \cap S^1$; we denote this new gap by $\hat{F}(G)$.

A gap G at depth $d > 0$ is *critical* if F is 2-to-1 on the interior of $G \cap S^1$. By definition, the unique largest gap at depth $d = 0$ is also critical.

Now let $\mathcal{P}_d$ denote the pieces at depth d in the Yoccoz puzzle for f. Since Λ_d separates the disk in the same pattern as the rays landing at $f^{-d}(\alpha)$ separate the plane, there is a natural depth-preserving bijection

$$\phi : \amalg \mathcal{P}_d \to \amalg \mathcal{G}_d$$

between the disjoint unions of the pieces and the gaps, such that:

1. an external ray R_t enters $P \in \mathcal{P}_d$ if and only if $\exp(2\pi i t) \in \phi(P) \in \mathcal{G}_d$;

2. the critical point $z = 0$ belongs to P if and only if $\phi(P)$ is a critical gap; and

3. if P is a puzzle piece at depth $d > 0$, then $\phi(f(P)) = \hat{F}(\phi(P))$.

The *model tableau* is constructed by setting $G_{d,0}$ equal to the unique critical gap at depth d, and defining $G_{d,k} = \hat{F}^k(G_{d+k,0})$. The model tableau is canonically determined by $\lambda_\alpha(f)$.

It is immediate that

$$
\begin{aligned}
\phi(P_d(c_k)) &= \phi(P_d(f^k(0))) = \phi(f^k(P_{d+k}(0))) = \hat{F}^k(G_{d+k,0}) \\
&= G_{d,k},
\end{aligned}
$$

so we have established:

Theorem 8.10 *There is a natural bijection* ϕ *between puzzle pieces and gaps which sends the critical piece* $P_d(c_k)$ *to the gap* $G_{d,k}$.

Example. Let $f(z) = z^2 + i$. The hyperbolic convex hulls of the equivalence classes of $\lambda_\alpha^4(f)$ are depicted in Figure 8.2; the puzzle pieces of depth four or less for the same map are shown in Figure 8.3.

Figure 8.2. The α-lamination for $z^2 + i$, to depth four.

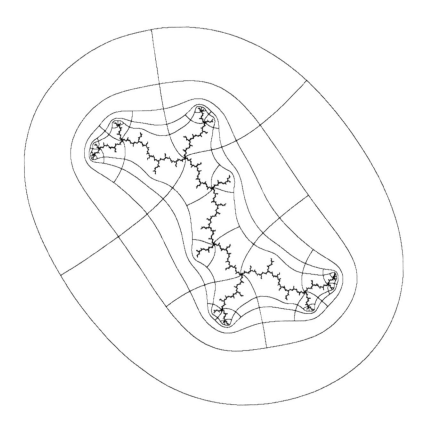

Figure 8.3. The puzzle pieces for $z^2 + i$, to depth four.

Definition. The model tableau is *periodic* if for some $n > 0$, $G_{d,0} = G_{d,n}$ for all depths d. The least such n is the *period of the model tableau*.

By Theorem 8.3 we have:

Corollary 8.11 *The period of the model tableau $G_{d,k}$ is equal to the renormalization period of f.*

Theorem 8.12 *Let f and g be quadratic polynomials, each with connected Julia set, both fixed points repelling, and the forward orbit of the critical point disjoint from the α fixed point.*

Suppose $\lambda_\alpha(f) \subset \lambda_\alpha(g)$. Then $\lambda_\alpha(f) = \lambda_\alpha(g)$ and the renormalization period of f is equal to that of g.

Proof. Suppose $\lambda_\alpha(f) \subset \lambda_\alpha(g)$. By Theorem 7.9, the external rays landing at the α fixed point are permuted transitively. In particular, the external angle of one ray landing at α determines the entire set of rays landing at α. Since $\lambda^0_\alpha(f) \subset \lambda^0_\alpha(g)$, and the former equivalence relation is nontrivial, we have $\lambda^0_\alpha(f) = \lambda^0_\alpha(g)$. Since the forward orbit of the critical point avoids α, the cardinalities of $f^{-d}(\alpha)$ and $g^{-d}(\alpha)$ are the same $(2^{d+1} - 1)$, and so $\lambda^d_\alpha(f)$ and $\lambda^d_\alpha(g)$ have the same number of nontrivial equivalence classes. Thus $\lambda^d_\alpha(f) \subset \lambda^d_\alpha(g)$ implies $\lambda^d_\alpha(f) = \lambda^d_\alpha(g)$ and therefore $\lambda_\alpha(f) = \lambda_\alpha(g)$.

Since λ_α determines the model tableau, the preceding result shows f and g have the same renormalization period.

∎

Remarks. This theorem fails if we do not require both fixed to be repelling. For example, if we set $f(z) = z^2 - 1$ and $g(z) = z^2 - 3/4$, then $f^2(z)$ is renormalizable and $\lambda_\alpha(f) = \lambda_\alpha(g)$, but $g^2(z)$ is not renormalizable because its α fixed point is parabolic.

Also, if we allow the critical point to land on α, then $\lambda_\alpha(f) \subset \lambda_\alpha(g)$ does not imply $\lambda_\alpha(f) = \lambda_\alpha(g)$. An example is provided by taking $f(z) = z^2 - 1$ and $g(z) = z^2 - 1.54368...$, where $g^3(0)$ is the α-fixed point of g. (Here g is the same as Example III of §7.4.)

Chapter 9

Robustness

In this chapter we turn to the geometric aspects of infinitely renormalizable polynomials. First, we define a canonical set of disjoint simple geodesics on the Riemann surface $\widehat{\mathbb{C}} - P(f)$. The disks bounded by these geodesics are like the basic intervals in the construction of a Cantor set. When these geodesics have length bounded above at infinitely many levels, we say the polynomial is *robust*. We show that the postcritical set of a robust polynomial is a Cantor set of zero area, and give a topological model for its dynamics.

9.1 Simple loops around the postcritical set

Definitions. Let f be infinitely renormalizable. Then $|\mathcal{SR}(f)| = \infty$ by Theorem 8.4. Let $\mathcal{SR}(f)^* = \mathcal{SR}(f) - \{1\}$.

For each level n in $\mathcal{SR}(f)^*$, let $\gamma_n(i)$ denote the hyperbolic geodesic on $\widehat{\mathbb{C}} - P(f)$ representing a simple closed curve separating $J_n(i)$ from $P(f) - J_n(i)$. This curve exists because $P_n(j)$ is disjoint from $J_n(i)$ for any $i \neq j$ (Theorem 8.1), its homotopy class is unique because $J_n(i)$ is connected, and it is represented by a geodesic because $n > 1$.

Let $\gamma_n = \gamma_n(n)$ denote the loop encircling the critical point of f.

The terminology "simple renormalization" is motivated by the following:

Theorem 9.1 (Simple curves) *The geodesics*

$$\Gamma = \{\gamma_n(i) : n \in \mathcal{SR}(f)^* \text{ and } 1 \leq i \leq n\}$$

are simple and disjoint.

Proof. A hyperbolic geodesic is simple if there is a simple representative in its homotopy class, and two distinct geodesics are disjoint if they are homotopic to disjoint curves. Thus $\gamma_n(i)$ is simple by definition.

Now consider two geodesics $\gamma_a(i)$ and $\gamma_b(j)$. We can find representatives of these curves in $\widehat{\mathbb{C}} - P(f)$ which are arbitrarily close to $J_a(i)$ and $J_b(j)$. If these small Julia sets are disjoint, then so are $\gamma_a(i)$ and $\gamma_b(j)$.

Otherwise $J_a(i)$ meets $J_b(j)$. If $a = b$, then $J_a(i) \cap J_b(j)$ is a single point x, a periodic point of f which does not belong to the postcritical set. *Since the renormalization is simple*, x does not disconnect $J_a(i)$ or $J_a(j)$. Thus we can represent $\gamma_a(i)$ and $\gamma_b(j)$ by homotopic curves which are disjoint, so the geodesic representatives are also disjoint.

Finally suppose $a < b$. Then $J_b(j) \subset J_a(k)$ for some k. It is obvious that $\gamma_b(j)$ can be represented by a curve nested inside $\gamma_a(k)$ in this case. Since the geodesics at level a bound disjoint disks, we need only rule out the possibility that $k = i$ and $\gamma_a(i) = \gamma_b(j)$. But there is a portion P' of the postcritical set lying in $J_a(i) - J_b(j)$, so $\gamma_a(i)$ and $\gamma_b(j)$ lie in different homotopy classes on $\widehat{\mathbb{C}} - P(f)$, and thus they are distinct and disjoint.

∎

Theorem 9.2 (Invariant curve system) *Let n belong to $\mathcal{SR}(f)^*$.*
For $i \neq 1$, $f^{-1}(\gamma_n(i))$ has a component α which is isotopic to $\gamma_n(i - 1)$ on $\widehat{\mathbb{C}} - P(f)$ and covers $\gamma_n(i)$ by degree one. The other component β bounds a disk disjoint from the postcritical set.
For $i = 1$, $f^{-1}(\gamma_n(1)) = \alpha$ is a connected curve isotopic to $\gamma_n(n)$ and covering $\gamma_n(1)$ by degree two.

Proof. The curve $\gamma_n(i)$ is isotopic on $\widehat{\mathbb{C}} - P(f)$ to a loop separating $J_n(i)$ from $P(f) - J_n(i)$. For $i \neq 1$ the critical value does not lie in $J_n(i)$, so the preimage of this loop has two components. One of them, α, separates $J_n(i-1)$ from $P(f) - J_n(i-1)$, and therefore is isotopic to $\gamma_n(i-1)$. The other component β is isotopic to a loop in a small neighborhood of $J'_n(i-1)$. Since $J'_n(i-1)$ does not meet the postcritical set, β bounds a disk on $\widehat{\mathbb{C}} - P(f)$.

When $i = 1$, the critical value lies in $J_n(1)$, so $f^{-1}(\gamma_n(1)) = \alpha$ covers $\gamma_n(1)$ by degree two and α is isotopic to a curve enclosing $J_n(n)$.

$\blacksquare$

Remark. As a consequence, the set of homotopy classes

$$\Gamma_n = \{\gamma_n(1), \ldots, \gamma_n(n)\}$$

on $\widehat{\mathbb{C}} - P(f)$ is an f-*invariant curve system*; see §B.2.

Let $\ell(\cdot)$ denote length in the hyperbolic metric on $\widehat{\mathbb{C}} - P(f)$.

Theorem 9.3 (Comparable lengths) *For any n in $\mathcal{SR}(f)^*$, the hyperbolic lengths of the geodesics $\gamma_n(i)$ at a given level n are comparable; in fact*

$$\frac{1}{2}\ell(\gamma_n(n)) \;\leq\; \ell(\gamma_n(1)) \;\leq\; \cdots \;\leq\; \ell(\gamma_n(n-1)) \;\leq\; \ell(\gamma_n(n)).$$

Proof. Let $Q = f^{-1}(P(f))$. For $1 < i \leq n$, the preimage $f^{-1}(\gamma_n(i))$ has a component α isotopic to $\gamma_n(i-1)$ and covering $\gamma_n(i)$ by degree one. Since $f : (\widehat{\mathbb{C}}-Q) \to (\widehat{\mathbb{C}}-P)$ is a covering, it is an isometry for the respective Poincaré metrics, while the inclusion $(\widehat{\mathbb{C}} - Q) \hookrightarrow (\widehat{\mathbb{C}} - P)$ is a contraction, so

$$\ell(\gamma_n(i-1)) \leq \ell(\alpha) \leq \ell(\gamma_n(i)).$$

Finally $f^{-1}(\gamma_n(1)) = \alpha$ is isotopic to $\gamma_n(n)$ and covers $\gamma_n(1)$ by degree two, which implies $\ell(\gamma_n(n)) \leq 2\ell(\gamma_n(1))$.

$\blacksquare$

9.2   Area of the postcritical set

Definition. A quadratic polynomial $f(z) = z^2 + c$ is *robust* if f is infinitely renormalizable, and

$$\liminf_{\mathcal{SR}(f)^*} \ell(\gamma_n) < \infty,$$

where $\ell(\cdot)$ denotes hyperbolic length on $\widehat{\mathbb{C}} - P(f)$.

Note that robustness is a property of the geometry of the postcritical set. The small Julia sets enter only in a combinatorial fashion, to pick out the simple closed curves γ_n.

The main result of this section is:

Theorem 9.4 (Postcritical measure zero) *Let f be robust. Then:*

1. *The postcritical set $P(f)$ is a Cantor set of measure zero.*

2. *As $n \to \infty$ in $\mathcal{SR}(f)$,*

$$\sup_i \operatorname{diam} P_n(i) \to 0.$$

3. *$f : P(f) \to P(f)$ is a homeomorphism, which is topologically conjugate to the map $x \mapsto x + 1$ acting on the group*

$$\operatorname*{proj\,lim}_{n \in \mathcal{SR}(f)} \mathbb{Z}/n.$$

The proof is based on the Collar Theorem and the area zero criterion of §2.8. So we actually obtain the stronger result that $P(f)$ has absolute area zero.

Proof. Let $\mathcal{SR}(f)^* = \{n(1), n(2), n(3), \dots\}$ with $n(k) < n(k+1)$.

For each k, and for $i = 1, \dots, n(k)$, let $A_k(i)$ be the standard collar $C(\gamma_{n(k)}(i))$ about the geodesic $\gamma_{n(k)}(i)$ on the hyperbolic surface $\widehat{\mathbb{C}} - P(f)$, as defined in §2.9. By the Collar Theorem 2.18, the collection of annuli obtained in this way are disjoint. Note that $A_k(i)$ separates $P_{n(k)}(i)$ from the rest of the postcritical set.

Let E_k be the union of the annuli around the small postcritical sets at level $n(k)$; that is,

$$E_k = \bigcup_{i=1}^{n(k)} A_k(i).$$

By Corollary 7.16, any small postcritical set $P_{n(k+1)}(i)$ is contained in $P_{n(k)}(j)$ for some j; therefore any annulus $A_{k+1}(i)$ in E_{k+1} is nested inside some annulus $A_k(j)$ in E_k.

We are assuming $\liminf \ell(\gamma_n) < \infty$. By Theorem 9.3, the geodesics at a given level $n(k)$ have length bounded above by $\ell(\gamma_{n(k)})$. Therefore the sum of the moduli of all annuli nested around an arbitrary point $x \in P(f)$ diverges, by Theorem 2.19. Applying Theorem 2.16, we conclude that the nested intersection $F = \bigcap F_k$ is a totally disconnected set of measure zero, where F_k is the union of the bounded components of $\mathbb{C} - E_k$. But each component of F_k meets $P(f)$, so $F = P(f)$ and we have shown that the postcritical set has measure zero.

Note that F_k is a decreasing sequence of compact sets and $P_{n(k)}(i)$ lies in a single component of F_k for any i. Since F is totally disconnected, the diameter of the largest component of F_k tends to zero, so $\sup_i \operatorname{diam}(P_{n(k)}(i))$ tends to zero as well.

For each $n \in \mathcal{SR}(f)$, there is a natural map $P(f) \to \mathbb{Z}/n$ sending $P_n(i)$ to $i \bmod n$. These maps are compatible as n varies, so they determine a continuous map

$$\phi : P(f) \to \operatorname*{proj\,lim}_{n \in \mathcal{SR}(f)} \mathbb{Z}/n.$$

We have just seen that the small postcritical sets nest down to points, so this map is injective; and it conjugates f to $x \mapsto x + 1$ because $f(P_n(i)) = P_n(i+1)$. The projectively limit is a Cantor set, so $P(f)$ is also a Cantor set.

∎

Remarks. The theorem also holds under the weaker hypothesis

$$\sum_{\mathcal{SR}(f)^\bullet} \bmod C(\gamma_n) = \infty,$$

where $C(\cdot)$ denotes the collar about γ_n.

It seems unlikely that every infinitely renormalizable quadratic polynomial is robust. A natural test case is provided by an example of Douady and Hubbard, where $f(z) = z^2 + c$ is infinitely renormalizable but the Julia set $J(f)$ is not locally connected [Mil3, §3].

We do not know if robustness implies the Julia set is locally connected, or has measure zero. Robustness seems to give more control over the postcritical set than the Julia set. Of course it would also be interesting to know if the Mandelbrot set is locally connected at c when $z^2 + c$ is robust.

Chapter 10

Limits of renormalization

This chapter is devoted to the proof of our main result, which we restate:

Theorem 1.7 (Robust rigidity) *A robust infinitely renormalizable quadratic polynomial f carries no invariant line field on its Julia set.*

This theorem is equivalent to the statement that any quadratic polynomial g which is quasiconformally conjugate to f is conformally conjugate to f. Because the theorem rules out quasiconformal deformations, we call it a rigidity result.

Recall that an infinitely renormalizable quadratic polynomial f is robust if

$$\liminf_{\mathcal{SR}(f)^\bullet} \ell(\gamma_n) = L < \infty.$$

We will divide the proof into two cases, depending on whether or not $L = 0$.

Both arguments follow the same pattern, which we now briefly summarize. Suppose f is infinitely renormalizable, robust, and carries an invariant line field μ on its Julia set. To deduce a contradiction, we pass to a subsequence of n in $\mathcal{SR}(f)$ such that after rescaling, f^n converges on a neighborhood of the small postcritical P_n. The limiting dynamical system f_∞ will be a proper map of degree two. Robustness furnishes the compactness required in this step.

Next we use μ to construct a *univalent* invariant line field for f_∞. Pick a point x in the Julia set of f where μ is almost continuous, such that $\|(f^k)'x\| \to \infty$ with respect to the hyperbolic metric on

141

$\widehat{\mathbb{C}} - P(f)$ as $k \to \infty$, and such that $f^k(x)$ tends to (but does not land in) $P(f)$ as $k \to \infty$. We would like to push μ forward univalently from a small neighborhood of x to a definite neighborhood of P_n.

The idea for obtaining a univalent pushforward is summed up in Figure 10.1. The circles labeled $1, 2, \ldots, n$ denote neighborhoods of the small postcritical sets $P_n(1)$, $P_n(2)$, $\ldots$, $P_n(n)$. (The exact choice of neighborhood of $P_n(i)$ varies between the cases $L = 0$ and $L > 0$; in the simpler case where $L = 0$, $V_n(i)$ will do.) For $i = 1, \ldots, n-1$, region i maps to region $i + 1$ univalently. Region n contains the critical point and maps over $P_n(1)$ by degree two (it may or may not map over region 1). Each region other than 1 has a primed companion with the same image.

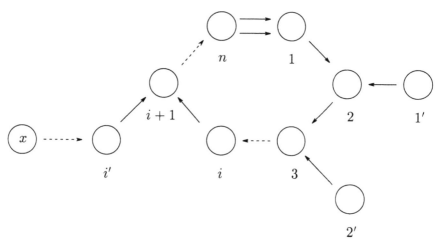

Figure 10.1. Pattern of the proof: univalent pushforward from x.

Since x is attracted to the postcritical set, it eventually enters the cycle of regions at level n. Just before it does so, it must enter a primed region i'. This region is disjoint from the postcritical set, so it admits a univalent pushforward from a neighborhood of x. There is sufficient expansion in the hyperbolic metric on $\widehat{\mathbb{C}} - P(f)$ to guarantee that this pushforward of μ is nearly univalent. The map f^{n-i} carries this line field around to a neighborhood of P_n. In the limit we obtain a univalent invariant line field for f_∞. But this is impossible, because f_∞ has a critical point. Thus the original dynamical system f has no invariant line field either.

When $L = 0$ (or indeed when L is sufficiently small), the limiting dynamical system f_∞ can be chosen to be a quadratic-like map (rather than just a proper map). This is the simplest situation to analyze, and it is treated in §10.1 and §10.2.

When $L > 0$, we can still construct a limiting proper map of degree two, but it may not be quadratic-like. In the course of the construction, we also cannot assume that the preimage of region 1 is contained in region n. Thus there is an additional possibility to analyze, when x lands in between. The new features which arise in this case are dealt with in §10.3 and §10.4.

10.1 Unbranched renormalization

Let f be a quadratic polynomial which admits an infinite sequence of simple renormalizations.

Definition. A renormalization $f^n : U_n \to V_n$ is *unbranched* if

$$V_n \cap P(f) = P_n.$$

Theorem 10.1 *If $f^n : U_n \to V_n$ is an unbranched renormalization, then $V_n'(i)$ is disjoint from the postcritical set of f for $i \neq n$.*

Proof. The small postcritical sets are disjoint and $f(P_n(i)) \subset P_n(i+1)$, so when $f^n : U_n \to V_n$ is unbranched we have $V_n(i) \cap P(f) = P_n(i)$. Since $V_n(i)$ and $V_n'(i)$ have the same image under f, any point in $V_n'(i) \cap P(f)$ must lie in $P_n(i)$. But $V_n(i)$ and $V_n'(i)$ are disjoint for $i \neq n$.

■

Therefore f^{-k} can be defined univalently on $V_n'(i)$ for any $k > 0$. The existence of univalent inverses is the reason for the terminology "unbranched".

In the next section, we will establish:

Theorem 10.2 (Polynomial-like rigidity) *Let f be a quadratic polynomial and let $m > 0$ be a constant. Suppose for infinitely many $n > 1$ there is a simple unbranched renormalization $f^n : U_n \to V_n$ with $\mathrm{mod}(U_n, V_n) > m > 0$.*

Then the Julia set of f carries no invariant line field.

Assuming this theorem for the moment, we can deduce the main rigidity result in the case where $\liminf \ell(\gamma_n)$ is sufficiently small. To do this, we must relate the length of γ_n to the existence of unbranched renormalizations.

Theorem 10.3 *Suppose f is infinitely renormalizable and f^n is simply renormalizable. Then we may choose U_n and V_n such that the renormalization $f^n : U_n \to V_n$ is unbranched. When $\ell(\gamma_n)$ is sufficiently small, we can also ensure*

$$\mathrm{mod}(U_n, V_n) > m(\ell(\gamma_n)) > 0,$$

where $m(\ell) \to \infty$ as $\ell \to 0$.

Proof. We can always choose V_n very close to J_n, by starting with an arbitrary renormalization and replacing V_n with a component of $f^{-nd}(V_n)$ for large d. Since $P(f) - K_n$ is closed, this renormalization is unbranched when V_n is sufficiently close to K_n.

To complete the proof, we will show that when γ_n is a short geodesic on $\widehat{\mathbb{C}} - P(f)$, there is an unbranched renormalization with $\mathrm{mod}(U_n, V_n)$ large.

Let A_n be the collar $C(\gamma_n)$ about γ_n with respect to the hyperbolic metric on $\widehat{\mathbb{C}} - P(f)$. The modulus of A_n is large when $\ell(\gamma_n)$ is small (cf. Theorems 2.18 and 2.19.) Let B_n be the component of $f^{-n}(A_n)$ which lies in the same homotopy class as γ_n. Construct open disks D_n and E_n by adjoining to A_n and B_n the bounded components of their complements; then $f^n : D_n \to E_n$ is a proper map of degree two, and

$$D_n \cap P(f) = E_n \cap P(f) = P_n$$

so $f^n : D_n \to E_n$ is a critically compact proper map (in the sense of §5.5.)

Since f is infinitely renormalizable, it has no attracting cycles. By Theorem 5.12, we can find a renormalization $f^n : U_n \to V_n$ with $U_n \subset D_n$ and $V_n \subset E_n$ whenever $\mathrm{mod}(P_n, E_n)$ is sufficiently large. Moreover, the renormalization can be constructed so that $\mathrm{mod}(U_n, V_n)$ is bounded below in terms of $\mathrm{mod}(P_n, E_n)$. Clearly

$$\mathrm{mod}(P_n, E_n) \geq \mathrm{mod}(A_n) = \mathrm{mod}(C(\gamma_n)),$$

and the modulus of the collar depends only on $\ell(\gamma_n)$ and tends to infinity as $\ell(\gamma_n)$ tends to zero. Thus $\mathrm{mod}(U_n, V_n)$ is bounded below in terms of $\ell(\gamma_n)$, and the theorem follows.

∎

Corollary 10.4 (Thin rigidity) *There is a constant $L > 0$ such that if*

$$\liminf_{SR(f)^\bullet} \ell(\gamma_n) < L,$$

then f carries no invariant line field on its Julia set.

In particular, the theorem holds if $\liminf \ell(\gamma_n) = 0$. This condition implies the Riemann surface $\hat{\mathbb{C}} - P(f)$ has infinitely many very thin parts.

Proof. By the previous theorem, there exists an $L > 0$ such that f^n admits an unbranched renormalization with $\mathrm{mod}(U_n, V_n) > 1$ whenever $\ell(\gamma_n) < L$. So the corollary follows by Theorem 10.2.

∎

The following corollary is not used in the sequel, but it clarifies the picture of thin rigidity. It also shows that many infinitely renormalizable maps — such as the Feigenbaum polynomial — do *not* satisfy $\liminf_{SR(f)^\bullet} \ell(\gamma_n) = 0$. In §10.3 we will prove rigidity for robust polynomials with $\liminf_{SR(f)^\bullet} \ell(\gamma_n) > 0$.

Corollary 10.5 *There is a constant $L' > 0$ such that for any infinitely renormalizable quadratic polynomial f, and any $n \in SR(f)^*$, $\ell(\gamma_n) < L'$ implies the renormalization of f^n is of disjoint type (the small Julia sets at level n do not touch).*

Proof. Suppose $\ell(\gamma_n)$ is small. Then the collar lemma provides a large annulus separating $P_n(i)$ from $P(f) - P_n(i)$ for each i. This annulus contains a round annulus of comparable modulus (Theorem 2.1), so

$$\operatorname{diam} P_n(i) \ll d(P_n(i), P(f) - P_n(i))$$

in the Euclidean metric. By Theorem 10.3, the polynomial-like mappings $f^n : U_n(i) \to V_n(i)$ can be chosen with $\operatorname{mod}(U_n, V_n)$ large as well. Since f^n is infinitely renormalizable, it has no attracting cycles, so by Corollary 5.10

$$\operatorname{diam}(K_n(i)) = O(\operatorname{diam}(P_n(i))).$$

Combining these two estimates, we conclude the sets $K_n(i)$ are disjoint when $\ell(\gamma_n)$ is sufficiently small. In fact

$$\frac{d(K_n(i), K_n(j))}{\operatorname{diam} K_n(i)} \geq C(\ell(\gamma_n))$$

where $C(\ell) \to \infty$ as $\ell \to 0$.

■

10.2   Polynomial-like limits of renormalization

In this section we prove Theorem 10.2 (Polynomial-like rigidity). First we obtain some preliminary information about unbranched renormalizations with definite moduli. This will enable us to construct a suitable point x from which to push the line field forward. Then we will pass to a limit (using compactness of quadratic-like maps with $\operatorname{mod}(U, V) > m$) and obtain a contradiction.

Theorem 10.6 *Suppose the renormalization of f^n is unbranched and $\operatorname{mod}(U_n, V_n) > m > 0$. Then for $i \neq n$, the companion Julia set satisfies*

$$\operatorname{diam}(J_n'(i)) \leq C(m)$$

in the hyperbolic metric on $\widehat{\mathbb{C}} - P(f)$.

Proof. By Theorem 2.4, a lower bound on the modulus of an annulus surrounding U_n in V_n gives an upper bound on the diameter of U_n in the hyperbolic metric on V_n. Since $J_n \subset V_n$, the hyperbolic diameter of J_n is also bounded by a constant $C(m)$ depending only on m. By Theorem 7.2, the pair $(J_n'(i), V_n'(i))$ is conformally isomorphic to the pair (J_n, V_n), so $C(m)$ bounds the hyperbolic diameter $J_n'(i)$ in $V_n'(i)$ as well. Since the renormalization is unbranched, $V_n'(i)$ is disjoint from the postcritical set of f; the inclusion $V_n'(i) \hookrightarrow \widehat{\mathbb{C}} - P(f)$ is a contraction (by the Schwarz lemma), so $C(m)$ also bounds the diameter of $J_n'(i)$ in the hyperbolic metric on $\widehat{\mathbb{C}} - P(f)$.

∎

Theorem 10.7 *If f has infinitely many unbranched simple renormalizations with*
$$\mathrm{mod}(U_n, V_n) > m > 0,$$
then f is robust, its finite postcritical set is a Cantor set of measure zero and
$$P(f) \cap \mathbb{C} = \bigcap_{SR(f)} J_n.$$

Proof. Suppose f^n is simply renormalizable and $\mathrm{mod}(U_n, V_n) > m$. Then the core curve of the annular component A_n of $V_n - \overline{U_n}$ has length $\pi/\mathrm{mod}(A_n) \le \pi/m$ in the hyperbolic metric on A_n (see §2.2). The core curve of A_n is homotopic to γ_n, so $\liminf_{SR(f)} \cdot \ell(\gamma_n)^{\cdot} \le \pi/m < \infty$ and therefore f is robust. By Theorem 9.4, the postcritical set is a Cantor set of measure zero. Moreover, $\sup_i \mathrm{diam}\, P_n(i) \to 0$ by the same Theorem.

Fixing any n with $\mathrm{mod}(U_n, V_n) > m$, we have
$$\mathrm{diam}\, J_n(i) \le C(m)\,\mathrm{diam}\, P_n(i)$$
by Corollary 5.10, since the renormalization $f^n : U_n(i) \to V_n(i)$ is conjugate to $f^n : U_n \to V_n$ and has no attracting fixed point.

Therefore $\sup_i \mathrm{diam}(J_n(i)) \to 0$ as $n \to \infty$ in $SR(f)$, which implies the nested intersection $\bigcap J_n$ is totally disconnected. Since each component of J_n meets the postcritical set, we have $P(f) \cap \mathbb{C} = \bigcap J_n$.

∎

Corollary 10.8 *Under the same hypotheses, almost every point x in $J(f)$ has the following properties:*

1. *the forward orbit of x does not meet the postcritical set;*

2. *$\|(f^n)'(x)\| \to \infty$ in the Poincaré metric on $\widehat{\mathbb{C}} - P(f)$;*

3. *for any $n \in \mathcal{SR}(f)$, there is a $k > 0$ such that $f^k(x) \in \mathcal{J}_n$; and*

4. *for any $k > 0$, there is an $n \in \mathcal{SR}(f)$ such that $f^k(x)$ does not meet $\mathcal{J}_n$.*

Proof. By the preceding Theorem, $P(f)$ has measure zero; therefore $\bigcup f^{-n}(P(f))$ has measure zero, establishing point one. The second and third points follow by Theorems 3.6 and 8.2.

To verify the last property, note that by the preceding Theorem, area($\mathcal{J}_n$) $= 0$ as $n \to \infty$ in $\mathcal{SR}(f)$. Fixing k, we have area($f^{-k}(\mathcal{J}_n)$) $\to 0$ as well, and therefore $\bigcap_n f^{-k}(\mathcal{J}_n)$ has measure zero.

∎

Proof of Theorem 10.2 (Polynomial-like rigidity).   Let $\mathcal{USR}(f, m)$ be an infinite set of n such that $f^n : U_n \to V_n$ is an unbranched simple renormalization and $\mathrm{mod}(U_n, V_n) > m$.

Suppose f admits an invariant line field μ supported on a set E of positive measure in the Julia set. Let $x \in E$ be a point of almost continuity of μ which also satisfies the properties guaranteed by the Corollary above. For each $n \in \mathcal{SR}(f)$, let $k(n) \geq 0$ be the least non-negative integer such that $f^{k(n)+1}(x) \in \mathcal{J}_n$. By our choice of x, $k(n) \to \infty$.

Consider $n \in \mathcal{USR}(f, m)$ sufficiently large that $k(n) > 0$. Then $f^{k(n)+1}(x) \in \mathcal{J}_n$ but $f^{k(n)}(x)$ is not in $\mathcal{J}_n$. Therefore $f^{k(n)}(x)$ lies in some companion Julia set $J_n'(i(n))$ where $0 < i(n) < n$. (Note $f^{k(n)+1}(x)$ cannot lie in $J_n(1)$ since that would imply $f^{k(n)}(x) \in J_n(n) \subset \mathcal{J}_n$.)

Since the nth renormalization is unbranched, there is a univalent branch of $f^{-k(n)}$ defined on $V_n'(i(n))$ and carrying $f^{k(n)}(x)$ back to x.

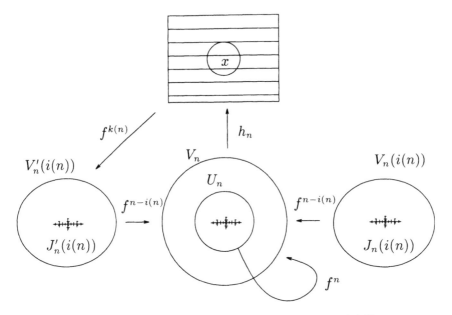

Figure 10.2. Expanding a line field into a polynomial-like map.

The map $f^{n-i(n)} : V'_n(i(n)) \to V_n$ is also univalent. Define a univalent map $h_n : V_n \to \mathbb{C}$ by the composition

$$V_n \xrightarrow{f^{i(n)-n}} V'_n(i(n)) \xrightarrow{f^{-k(n)}} \mathbb{C}$$

using the branches of the inverse mentioned above. See Figure 10.2. By invariance, $\mu | V_n = h_n^*(\mu)$.

Let

$$J_n^* \;=\; h_n(J_n) \;=\; f^{-k(n)}(J'_n(i(n)))$$

be the small copy of the Julia set containing x. Since the diameter of $J'_n(i(n))$ in the hyperbolic metric on $\hat{\mathbb{C}} - P(f)$ is bounded above (Theorem 10.6), and $\|(f^{k(n)})'(x)\| \to \infty$, we have

$$\mathrm{diam}(J_n^*) \to 0$$

in the hyperbolic metric on $\hat{\mathbb{C}} - P(f)$. (To show the derivative at one point controls the size of J_n^*, we may appeal to Theorem 3.8 or to the Koebe distortion theorem.)

Thus the line field $\mu = h_n^*(\mu)$ is invariant under f^n and close to a univalent line field on V_n.

Let $A_n(z) = z/\operatorname{diam}(J_n)$, let $g_n = A_n \circ f^n \circ A_n^{-1}$, and let $y_n = A_n(h_n^{-1}(x))$. Then

$$g_n : (A_n(U_n), 0) \to (A_n(V_n), A_n(f^n(0)))$$

is a quadratic-like map with $\operatorname{diam}(J(g_n)) = 1$, satisfying

$$\operatorname{mod}(A_n(U_n), A_n(V_n)) \geq m.$$

Since $h_n^{-1}(x) \in J_n$, we have $y_n \in J(g_n)$.

By Theorem 5.8, there is a subsequence of n in $\mathcal{USR}(f, m)$ such that g_n converges in the Carathéodory topology to a quadratic-like map

$$g : (U, 0) \to (V, g(0))$$

with $\operatorname{mod}(U, V) \geq m$.

Let $k_n = h_n \circ A_n^{-1}$ denote the composition

$$A_n(V_n) \xrightarrow{A_n^{-1}} V_n \xrightarrow{h_n} \mathbb{C}.$$

Then $k_n(y_n) = x$ and $\nu_n = k_n^*(\mu)$ is a g_n-invariant line field on $A_n(V_n)$.

Since $1 = \operatorname{diam}(J(g_n))$, while $\operatorname{diam}(k_n(J(g_n))) = \operatorname{diam}(J_n^*) \to 0$, we have $k_n'(y_n) \to 0$ by the Koebe theorem. The fact that $y_n \in J(g_n)$ and the Julia set is surrounded by an annulus of definite modulus in $A_n(V_n)$ implies for a further subsequence, $(A_n(V_n), y_n) \to (V, y)$ by Theorem 5.3. By Theorem 5.16, after passing to a further subsequence, ν_n converges to a *univalent* g-invariant line field ν on V.

As the renormalizations f^n have connected Julia sets, so does g; therefore the critical point and critical value of g lie in V. But then g admits no univalent invariant line field (by Theorem 5.13).

Therefore f itself has no measurable invariant line field.

∎

Remark. When $\liminf_{\mathcal{SR}(f)} \ell(\gamma_n) = 0$, the proof above takes a particularly simple form. In this case, the polynomial-like maps g_n can be chosen to converge to a quadratic *polynomial* $g : \mathbb{C} \to \mathbb{C}$, and the line fields ν_n to the family ν of horizontal lines in the plane. Clearly ν fails to be g-invariant; indeed the only polynomials preserving ν are the linear maps $az + b$ with $a \in \mathbb{R}$.

10.3 Proper limits of renormalization

In this section and the next we complete the proof of Theorem 1.7 (Robust rigidity) by establishing:

Theorem 10.9 (Thick rigidity) *Let* $f(z) = z^2 + c$ *be infinitely renormalizable, and suppose*

$$0 < \liminf_{\mathcal{SR}(f)^*} \ell(\gamma_n) < \infty.$$

Then f *carries no invariant line field on its Julia set.*

The proof follows the same outline as the proof in §10.2. There is one important difference: we do not use polynomial-like mappings. Indeed, we do not know if a robust quadratic polynomial admits infinitely many renormalizations with $\mathrm{mod}(U_n, V_n) > m > 0$.[1] Because of this we have no control over the shape of the small Julia sets.

This makes it more difficult to say when the forward orbit $\langle f^k(x) \rangle$ first enters the influence of a given level n of renormalization (whereas before we could simply look at the least k such that $f^k(x) \in J_n$). To take care of this, we will show that by the time $f^k(x)$ is quite close to a small postcritical set $P_n(i)$ at level n, it has already passed close to some companion postcritical set $P_n'(j)$ at the same level. In this way we will obtain a nearly univalent invariant line field near $P_n'(j)$, and from that a contradiction.

To carry out the proof, we will need infinitely many geodesics γ_n which are neither too long *nor* too short, to obtain a kind of bounded geometry. Thus the arguments in the thick and thin cases, while similar, are really distinct: neither one contains the other.

Definitions. We collect together some additional notation. For each $n \in \mathcal{SR}(f)^*$:

- δ_n denotes the unique component of $f^{-n}(\gamma_n)$ which is isotopic to γ_n on $\widehat{\mathbb{C}} - P(f)$. This curve exists by Theorem 9.2.

- X_n denotes the disk in $\mathbb{C}$ bounded by δ_n.

---

[1] Sullivan has obtained a proof of this property for real quadratics of *bounded type*; see [Sul4].

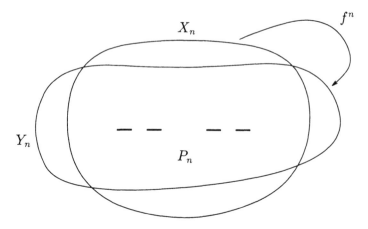

Figure 10.3. Proper map of degree two at level n.

- Y_n denotes the disk bounded by γ_n.

  Then $f^n : X_n \to Y_n$ is a proper map of degree two (see Figure 10.3).

- $Y_n(i) = f^i(X_n)$ for $i = 1, 2, \ldots n$. The map $f^n : X_n \to Y_n$ factors as

$$X_n \xrightarrow{f} Y_n(1) \xrightarrow{f} \ldots \xrightarrow{f} Y_n(n) = Y_n,$$

  where the first map $f : X_n \to Y_n(1)$ is proper of degree two and the remaining maps are univalent. Note that

$$Y_n(i) \cap P(f) = P_n(i).$$

- $\mathcal{Y}_n$ is the union $\bigcup_1^n Y_n(i)$. Then $P(f) \subset \mathcal{Y}_n$.

- $Y_n'(i) = -Y_n(i)$ for $i = 1, \ldots, n-1$. This disk is disjoint from $Y_n(i)$ (since each maps univalently to $Y_n(i+1)$), and therefore $Y_n'(i)$ is disjoint from the postcritical set $P(f)$.

- B_n denotes the largest Euclidean ball centered at the critical point $z = 0$ and contained in $X_n \cap Y_n$.

Theorem 10.10 *Let f be robust. Then*

$$P(f) \cap \mathbb{C} = \bigcap_{\mathcal{SR}(f)} \mathcal{Y}_n.$$

Proof. By Theorem 9.4, $\sup_i \operatorname{diam}(P_n(i)) \to 0$ as $n \to \infty$ in $\mathcal{SR}(f)$. Since $P_n(i) \subset Y_n(i)$, we just need to check that $\sup_i \operatorname{diam}(Y_n(i)) \to 0$ as well.

Choose λ so that $\ell(\gamma_a) < \lambda$ for infinitely many $a \in \mathcal{SR}(f)^*$. For each such a, the length of $\partial Y_a(i)$ is also bounded by λ in the hyperbolic metric on $\widehat{\mathbb{C}} - P(f)$. Now if $b < a$ in $\mathcal{SR}(f)^*$, then $\partial Y_a(i)$ separates $P_b(i)$ into two pieces (since $P_a(i)$ and $P_a(i+b)$ are both contained in $P_b(i)$). When b is large, the diameter of $P_b(i)$ is small, and so $Y_a(i)$ passes close to the postcritical set $P(f)$. Since the hyperbolic length of $\partial Y_a(i)$ is bounded, it follows that $\operatorname{diam}(Y_a(i))$ is also small.

■

Corollary 10.11 *Let f be robust. Then almost every point x in $J(f)$ has the following properties:*

1. *the orbit of x does not meet the postcritical set;*

2. *$\|(f^n)'(x)\| \to \infty$ in the Poincaré metric on $\widehat{\mathbb{C}} - P(f)$;*

3. *for any $n \in \mathcal{SR}(f)^*$, there is a $k > 0$ such that $f^k(x) \in \mathcal{Y}_n$; and*

4. *for any $k > 0$, there is an $n \in \mathcal{SR}(f)^*$ such that $f^k(x)$ does not meet $\mathcal{Y}_n$.*

Proof. These properties follow immediately from Theorems 9.4, 3.6, 3.9 and 10.10. Compare Corollary 10.8.

■

Definition. Let

$$SR(f, \lambda) \;=\; \{n \in SR(f)^* \;:\; 1/\lambda < \ell(\gamma_n) < \lambda\}.$$

When $0 < \liminf \ell(\gamma_n) < \infty$, the set $SR(f, \lambda)$ is infinite for some finite value of λ.

Theorem 10.12 *If $n \in SR(f, \lambda)$, then in the Euclidean metric,*

$$\begin{aligned}
\operatorname{diam}(X_n) &\geq \operatorname{diam}(B_n) \geq C(\lambda) \operatorname{diam}(X_n) \;\; and \\
\operatorname{diam}(Y_n) &\geq \operatorname{diam}(B_n) \geq C(\lambda) \operatorname{diam}(Y_n),
\end{aligned}$$

where $C(\lambda) > 0$.

Proof. Since $B_n \subset X_n \cap Y_n$, the inequalities on the left are trivial. For the inequalities on the right we use the existence of annuli of definite moduli around γ_n and δ_n.

First, $\ell(\gamma_n) < \lambda$ implies by the Collar Theorem 2.18 there is an annulus $A \subset \widehat{\mathbb{C}} - P(f)$ with core curve γ_n and with $\operatorname{mod}(A) > m(\lambda) > 0$. Since the critical point $z = 0$ belongs to $P(f)$, by Theorem 2.5 we have

$$r_1 = d(0, \gamma_n) \geq C_1 \operatorname{diam}(\gamma_n)$$

where $C_1 > 0$ depends only on λ.

As for δ_n, if we let $Q = f^{-n}(P(f))$ then the map

$$f^n : (\widehat{\mathbb{C}} - Q) \to (\widehat{\mathbb{C}} - P(f))$$

is a covering map sending δ_n to γ_n by degree two. Since covering maps are isometries for the hyperbolic metric, the length of δ_n on $\widehat{\mathbb{C}} - Q$ is at most 2λ. So by the same reasoning,

$$r_2 = d(0, \delta_n) \geq C_2 \operatorname{diam}(\delta_n)$$

where $C_2 > 0$ also depends only on λ.

Now we use the lower bound $\ell(\gamma_n)$ to show r_1 and r_2 are comparable. Suppose, for example, $r_1 \geq r_2/C_2 \geq \operatorname{diam}(\delta_n)$. Then the annulus $A = \{z \;:\; r_2/C_2 < |z| < r_1\}$ encloses δ_n and is enclosed by γ_n, so its core curve is homotopic to γ_n on $\widehat{\mathbb{C}} - P(f)$. By the Schwarz lemma, the length of the core curve of A in the hyperbolic metric on

A is bounded below by $\ell(\gamma_n) > 1/\lambda$. Using the formulas for modulus and length of the core curve given in §2, we have

$$\lambda \geq \frac{\mod(A)}{\pi} = \frac{\log r_1 - \log r_2 + \log C_2}{2\pi^2}.$$

Thus $r_1 \leq C_3 r_2$ for a constant C_3 depending only on λ. A similar argument bounds r_2 in terms of r_1. Thus the ball $B_n = \{z : |z| < \min(r_1, r_2)\}$ has diameter comparable to both $\mathrm{diam}(X_n)$ and $\mathrm{diam}(Y_n)$.

■

Theorem 10.13 *Suppose* $|\mathcal{SR}(f, \lambda)| = \infty$. *Define the affine map* A_n *by*

$$A_n(z) = \frac{z}{\mathrm{diam}(B_n)}.$$

Then in the Carathéodory topology, there is a subsequence of $n \in \mathcal{SR}(f, \lambda)$ *such that*

$$
\begin{aligned}
(A_n(X_n), 0) &\rightarrow (X, 0), \\
(A_n(Y_n), f^n(0)) &\rightarrow (Y, g(0)) \ \ and \\
A_n^{-1} \circ f^n \circ A_n &\rightarrow g,
\end{aligned}
$$

where $g : (X, 0) \rightarrow (Y, g(0))$ *is a proper map of degree two,* $0 \in X \cap Y$ *and* $g'(0) = 0$.

Proof. By our choice of A_n, the unit disk Δ is contained in both $A_n(X_n)$ and $A_n(Y_n)$. Therefore Theorem 5.2 gives a subsequence such that $(A_n(X_n), 0) \rightarrow (X, 0)$ and $A_n(Y_n), 0) \rightarrow (Y, 0)$.

The upper bound on $\ell(\gamma_n)$ provides an annulus of definite modulus between ∂Y_n and P_n, by the Collar Theorem 2.18. Thus the diameter of P_n in the hyperbolic metric on Y_n is bounded above independent of n; in particular, the hyperbolic distance $d(0, f^n(0))$ is bounded. By Theorem 5.3, after passing to a further subsequence, $(A_n(Y_n), f^n(0)) \rightarrow (Y, y)$.

By Theorem 10.12, the rescalings of X_n and Y_n have diameter bounded by $C(\lambda)$, so the limiting regions X and Y are not equal to $\mathbb{C}$.

By Theorem 5.6(Limits of proper maps), there is a further subsequence such that $A_n^{-1} \circ f^n \circ A_n$ converges to a proper map $g : X \to Y$ of degree at most two with $g(0) = y$. But $(f^n)'(0) = 0$, so $g'(0) = 0$ and therefore the degree of g is exactly two.

■

10.4 Extracting a univalent line field

Continuing with the notation of Theorem 10.13, we will next establish:

Theorem 10.14 (Proper invariant line field) *Suppose*

$$|\mathcal{SR}(f, \lambda)| = \infty$$

and f admits an invariant line field μ supported on its Julia set. Then there is a further subsequence such that $\mu_n = (A_n^{-1})^*(\mu|Y_n)$ converges to a univalent g-invariant line field ν on Y.

As in §10.2, the idea of the proof is to choose a point x of almost continuity of μ, push the line field forward and extract a limit. There are two cases to consider. In the first case, x lands in a companion disk $Y_n'(i)$ before it lands in $Y_n(i+1)$; since $Y_n'(i)$ is disjoint from the postcritical set, we may then push the line field forward *univalently* to $Y_n'(i)$, then to $Y_n(i+1)$, and then to $Y_n = Y_n(n)$. This case is the simplest and it follows the same lines as the proof in §10.2.

In the second case, x lands in $X_n - Y_n$ and then in $Y_n(1)$; in other words, it enters the disks $\bigcup Y_n(i)$ by first coming close to the critical point. Since X_n meets the postcritical set, we cannot necessarily construct a univalent branch from X_n back to a neighborhood of x. To handle this case, we will construct a disk Z_n close to X_n and *disjoint* from the postcritical set, which maps univalently to Y_n under f^m, where $0 < m < n$. Although x need not land in Z_n, it lands close enough that the line field spreads out to a nearly univalent line field on Z_n. By pushing forward from Z_n to Y_n we again obtain a univalent line field in the limit.

The construction of Z_n will use the complex shortest interval argument (§2.10). We will also apply Theorem 3.8 (Variation of expansion) to make several geometric bounds. Below, $C(\lambda)$ denotes a generic constant depending only on λ; different occurrences of $C(\lambda)$ are not meant to be the same. Distances and lengths, denoted $d(\cdot)$ and $\ell(\cdot)$, are measured in the hyperbolic metric on $\widehat{\mathbb{C}} - P(f)$.

Theorem 10.15 *For any $n \in \mathcal{SR}(f)^*$,*

$$
\begin{array}{ccccc}
\ell(\gamma_n) & \leq & \ell(\delta_n) & \leq & 2\ell(\gamma_n), \quad and \\
\ell(\gamma_n(i)) & \leq & \ell(\partial Y_n(i)) & \leq & \ell(\gamma_n)
\end{array}
$$

in the hyperbolic metric on $\widehat{\mathbb{C}} - P(f)$.

Proof. The left inequalities follow from the fact that γ_n and $\gamma_n(i)$ are the geodesic representatives of the isotopy classes δ_n and $\partial Y_n(i)$.

On the other hand, f^n maps δ_n to γ_n by degree two and f^{n-i} maps $\partial Y_n(i)$ to γ_n by degree one, so the inequalities on the right follow from the fact that f expands the hyperbolic metric on $\widehat{\mathbb{C}} - P(f)$.

∎

Corollary 10.16 *For any $n \in \mathcal{SR}(f, \lambda)$, δ_n is contained in a $C(\lambda)$-neighborhood of γ_n and $\partial Y_n(i)$ is contained in a $C(\lambda)$-neighborhood of $\gamma_n(i)$.*

Proof. The curve δ_n is isotopic to the geodesic γ_n on $\widehat{\mathbb{C}} - P(f)$. By the preceding theorem, the length of δ_n is bounded above by 2λ, while $n \in \mathcal{SR}(f, \lambda)$ implies the length of γ_n is bounded below by $1/\lambda$. Theorem 2.23 then guarantees every point in δ_n is within a bounded distance of γ_n.

A similar argument applies to $\partial Y_n(i)$, using the fact that $\ell(\gamma_n(i)) \geq \ell(\gamma_n)/2 > 1/(2\lambda)$ by Theorem 9.3.

∎

Theorem 10.17 *For any n in $\mathcal{SR}(f, \lambda)$,*

$$
\begin{array}{lll}
\|(f^n)'(x)\| & \leq & C(\lambda) \quad \text{for all } x \text{ in } \delta_n = \partial X_n; \text{ and} \\
\|(f^{n-i})'(x)\| & \leq & C(\lambda) \quad \text{for all } x \text{ in } \partial Y_n(i),
\end{array}
$$

with respect to the hyperbolic metric on $\widehat{\mathbb{C}} - P(f)$.

Proof. To prove the first inequality, note that f^n maps δ_n to γ_n by degree two, so

$$2\ell(\gamma_n) \;=\; \int_{\delta_n} \|(f^n)'(z)\| \, \rho(z)|dz|$$

where $\rho(z)|dz|$ denotes the hyperbolic metric on $\widehat{\mathbb{C}} - P(f)$. Since $\ell(\delta_n) \geq \ell(\gamma_n)$, we have $\|(f^n)'(x_1)\| \leq 2$ for some x_1 in δ_n.

Any point x_2 in δ_n is joined to x_1 by an arc η along δ_n, such that $f^n(\eta)$ is a subarc of γ_n; in particular, $\ell(f^n(\eta)) \leq \ell(\gamma_n) < \lambda$. By Theorem 3.8,

$$\|(f^n)'(x_2)\| \leq \|(f^n)'(x_1)\|^\alpha \leq 2^\alpha,$$

where α depends only on the hyperbolic length of $f^n(\eta)$, and hence is bounded in terms of λ.

Every x in $\partial Y_n(i)$ is equal to $f^i(x')$ for some x' in δ_n, so the second inequality follows from the first and the fact that f is an expansion.

∎

Theorem 10.18 *For any* $n \in \mathcal{SR}(f, \lambda)$, *there exist* $i \neq j$ *such that*

$$d(\gamma_n(i), \gamma_n(j)) < C(\lambda)$$

in the hyperbolic metric on $\widehat{\mathbb{C}} - P(f)$.

Proof. Let X' be the finite area hyperbolic surface obtained by deleting from the plane the open disks enclosed by the disjoint simple curves $\gamma_n(i)$. Then X' has geodesic boundary $\bigcup_i \gamma_n(i)$ and a single cusp at infinity. By Theorem 9.3, each boundary component has length bounded below by $L = \ell(\gamma_n)/2 \geq 1/(2\lambda)$. By Theorem 2.24, two boundary components $\gamma_n(i)$ and $\gamma_n(j)$ are within distance $D(L)$.

∎

Theorem 10.19 *For $n \in \mathcal{SR}(f, \lambda)$, there exists a disk $Z_n \subset \widehat{\mathbb{C}} - P(f)$ and an integer m, $0 < m < n$ such that*

1. *$f^m : Z_n \to Y_n$ is a univalent map;*

2. *$d(\partial X_n, \partial Z_n) < C(\lambda)$;*

3. *$\ell(\partial Z_n) < \lambda$; and*

4. *area$(Z_n) > 1/C(\lambda)$*

in the hyperbolic metric on $\widehat{\mathbb{C}} - P(f)$.

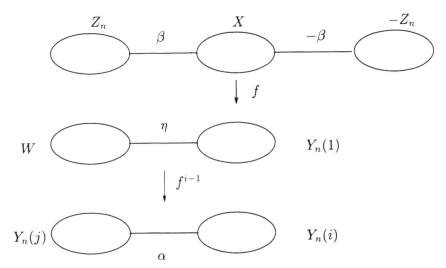

Figure 10.4. Construction of Z_n.

Remark. It may be the case Z_n meets X_n.

Proof. By the preceding result, there are $\gamma_n(i)$ and $\gamma_n(j)$ such that the hyperbolic distance $d(\gamma_n(i), \gamma_n(j))$ bounded in terms of λ. By Corollary 10.16, $Y_n(i)$ is uniformly close to $\gamma_n(i)$, so $d(Y_n(i), Y_n(j)) < C(\lambda)$ as well. We may assume $i < j$.

Let α be a geodesic of length less than $C(\lambda)$ joining $\partial Y_n(i)$ and $\partial Y_n(j)$. (If $Y_n(i)$ intersects $Y_n(j)$ we take α to be a geodesic of length zero.)

The map $f^{i-1} : Y_n(1) \to Y_n(i)$ is univalent, so it admits a univalent inverse

$$f^{1-i} : Y_n(i) \to Y_n(1).$$

Since $(\alpha \cup Y_n(j)) \cap P(f) = P_n(j)$, there is an analytic continuation of this branch of f^{1-i} to a univalent map sending $\alpha \cup Y_n(j)$ to $\eta \cup W$, where η is an arc joining the disk W to $Y_n(1)$. Note that W is either equal to $Y_n(j-i+1)$ or it is disjoint from the postcritical set.

Since $f(0)$ lies in $Y_n(1)$, the set

$$W \cup \eta \cup \partial Y_n(1)$$

has a two-fold cover under f equal to

$$Z_n \cup \beta \cup \partial X_n \cup -\beta \cup -Z_n,$$

where $f(Z_n) = f(-Z_n) = W$. (See Figure 10.4.)

After possibly replacing Z_n by $-Z_n$, we can assume Z_n is disjoint from the postcritical set. (Indeed, if Z_n meets $P(f)$, then $Z_n = Y_n(j-i)$ and $-Z_n = Y_n'(j-i)$ is disjoint from $P(f)$.)

Since $f^i : Z_n \to Y_n(j)$ and $f^{n-j} : Y_n(j) \to Y_n(n) = Y_n$ are univalent maps, so is $f^m : Z_n \to Y_n$, where $m = n+i-j < n$.

The map f^m is an expansion, so we have

$$\ell(\partial Z_n) \le \ell(\gamma_n) < \lambda.$$

Similarly, $\ell(\beta) \le \ell(\alpha)$, so $d(\partial Z_n, \partial X_n)$ is bounded in terms of λ.

Finally we show there is a lower bound on area(Z_n) depending only on λ. The idea is that f^m is not too expanding near ∂Z_n, and it maps a neighborhood of ∂Z_n to a region of definite area in Y_n.

To begin with, let

$$E_1 = \{z : d(z, \gamma_n) < 1\} \cap Y_n.$$

The Collar Theorem 2.18 provides a lower bound (depending on λ) for the injectivity radius of $\hat{\mathbb{C}} - P(f)$ along γ_n, and therefore a lower bound on area(E_1).

The map $f^{n-j} : Y_n(j) \to Y_n$ is univalent, so there is a region $E_2 \subset Y_n(j)$ mapping injectively to E_1. By Theorems 10.17 and 3.8, we have $\|(f^{n-j})'(z)\| < C(\lambda)$ in E_2, so we obtain a lower bound on area(E_2).

Similarly, the map $f^i : Z_n \to Y_n(j)$ carries a region $E_3 \subset Z_n$ injectively to E_2. Any point $z \in E_3$ can be joined to ∂X_n by an arc η such that $\ell(f^i(\eta)) < C(\lambda)$. Indeed, we may arrange that $f^i(\eta)$ consists of a geodesic joining $f^i(z)$ to $\partial Y_n(j)$, followed by an arc along $\partial Y_n(j)$, followed by β. Thus Theorem 3.8 bounds $\|(f^i)'(z)\|$ in terms of λ and the max of $\|(f^i)'\|$ along ∂X_n. But the latter is bounded in terms of λ by Theorem 10.17.

Since $f^i(E_3) = E_2$, we obtain a lower bound on $\mathrm{area}(E_3)$ in terms of λ, and hence a lower bound on $\mathrm{area}(Z_n)$.

■

Proof of Theorem 10.14 (Proper invariant line field). Let x be a point of almost continuity of μ, enjoying the properties guaranteed by Corollary 10.11 for almost every $x \in J(f)$.

For each $n \in \mathcal{SR}(f,\lambda)$, let $k(n) \geq 0$ be the least non-negative integer such that $f^{k(n)+1}(x) \in \mathcal{Y}_n$. Then $k(n) \to \infty$ as $n \to \infty$. We will only consider n large enough that $k(n) > 0$, so $f^{k(n)}(x)$ is *not* in $\mathcal{Y}_n$.

Next we construct univalent maps $h_n : Y_n \to T_n \subset \mathbb{C}$. We have $f^{k(n)+1}(x) \in Y_n(i(n)+1)$ for some $i(n)$ with $0 \leq i(n) \leq n-1$. Two cases will be distinguished, depending on whether $i(n) > 0$ or $i(n) = 0$.

Case I: $i(n) > 0$. Then $f^{k(n)}(x) \in Y_n'(i(n))$ (since $f^{k(n)}(x) \notin \mathcal{Y}_n$). Define $h_n : Y_n \to \mathbb{C}$ by the following composition:

$$ Y_n \xrightarrow{f^{i(n)-n}} Y_n'(i(n)) \xrightarrow{f^{-k(n)}} T_n \subset \mathbb{C}, $$

where the univalent branch of $f^{-k(n)}$ is chosen to send $f^{k(n)}(x)$ back to x. This branch is defined on all of $Y_n'(i(n))$ because $Y_n'(i(n))$ is disjoint from the postcritical set.

Case II: $i(n) = 0$. Then $f^{k(n)+1}(x) \in Y_n(1)$ but $f^{k(n)}(x) \notin Y_n(n) = Y_n$. Thus $f^{k(n)}(x) \in X_n - Y_n$.

Note that $X_n - Y_n$ is disjoint from the postcritical set, since ∂X_n and ∂Y_n are homotopic in $\widehat{\mathbb{C}} - P(f)$. By Corollary 10.16, ∂X_n is contained in a $C(\lambda)$ neighborhood of ∂Y_n, so the same is true of the whole region $X_n - Y_n$; in particular, the distance from $f^{k(n)}(x)$ to ∂X_n is so bounded. By Theorem 10.19, the distance from ∂X_n to

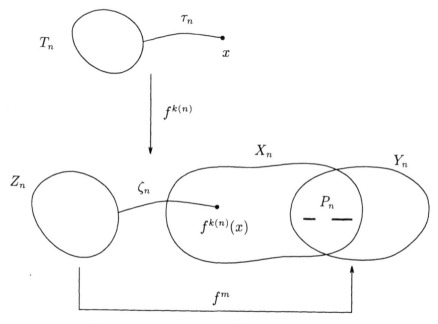

Figure 10.5. Case II: $f^{k(n)}(x) \in X_n - Y_n$.

Z_n is bounded in terms of λ, and the length of ∂X_n is bounded (by 2λ), so altogether we have

$$d(f^{k(n)}(x), Z_n) \leq C(\lambda).$$

Let ζ_n be a geodesic of minimal length joining $f^{k(n)}(x)$ to Z_n. Since $\zeta_n \cup Z_n$ is disjoint from the postcritical set, there is a univalent branch of $f^{-k(n)}$ defined on a neighborhood of this set, sending $f^{k(n)}(x)$ back to x and sending ζ_n to an arc connecting the image of Z_n to x.

By Theorem 10.19, there is a univalent map $f^m : Z_n \to Y_n$ for some $m > 0$. Then we define $h_n : Y_n \to T_n \subset \mathbb{C}$ as the composition

$$Y_n \xrightarrow{\ f^{-m}\ } Z_n \xrightarrow{\ f^{-k(n)}\ } T_n \subset \mathbb{C}.$$

See Figure 10.5.

Lemma 10.20 *As $n \to \infty$ in $\mathcal{SR}(f, \lambda)$,*

$$\operatorname{diam}(T_n) \to 0$$

while

$$d(x, T_n) \leq C(\lambda) \operatorname{diam}(T_n)$$

in the hyperbolic metric on $\widehat{\mathbb{C}} - P(f)$.

Proof. We will use the fact that $\|(f^{k(n)})'(x)\| \to \infty$ as $n \to \infty$.

In case I, $f^{k(n)}$ maps T_n univalently to $Y_n'(i(n))$, sending x into the interior of $Y_n'(i(n))$. Since the diameter of $\partial Y_n'(i(n))$ is bounded independent of n, we may apply Theorem 3.8 to conclude that $\|(f^{k(n)})'(y)\|$ tends to infinity uniformly for y in ∂T_n. Since the length of

$$\partial Y_n'(i(n)) = f^{k(n)}(\partial T_n)$$

is also bounded in terms of λ, we conclude that $\ell(\partial T_n)$ (and hence $\operatorname{diam}(T_n)$) tends to zero. The second assertion is trivial since $x \in T_n$.

In case II, $f^{k(n)}$ maps T_n univalently to Z_n. We claim

$$\frac{1}{C(\lambda)} \leq \frac{\|(f^{k(n)})'(y)\|}{\|(f^{k(n)})'(x)\|} \leq C(\lambda)$$

for all $y \in T_n \cup \tau_n$. First, the injectivity radius of $\widehat{\mathbb{C}} - P(f)$ is bounded below (in terms of λ) at $f^{k(n)}(x)$, since this point is within a bounded distance of the geodesic γ_n and $\ell(\gamma_n) > 1/\lambda$. (Here we use the Collar Theorem 2.18 and the fact that the log of the injectivity radius is Lipschitz, by Corollary 2.22.) Secondly, y can be joined to x by an arc η such that

$$\ell(f^{k(n)}(\eta)) \leq \ell(\zeta_n) + \operatorname{diam}(Z_n);$$

since the latter quantity is bounded in terms of λ, the claim follows from the second part of Theorem 3.8.

Since the arc η maps to an arc of bounded length, we have

$$d(x, y) \leq \frac{C(\lambda)}{\|(f^{k(n)})'(x)\|}$$

for any y in T_n. In particular this shows $\operatorname{diam}(T_n) \to 0$ in case II.

To check the last condition, we use the fact that the area of Z_n is bounded below in terms of λ (Theorem 10.19). Using the fact that the norm of the derivative varies by a bounded factor, we have

$$\begin{aligned} \frac{1}{C(\lambda)} &\leq \operatorname{area}(Z_n) = \int_{T_n} \|(f^{k(n)})'(y)\|^2 \, \rho(y)^2 |dy|^2 \\ &\leq C(\lambda) \operatorname{area}(T_n) \|(f^{k(n)})'(x)\|^2, \end{aligned}$$

where $\rho(y)^2 |dy|^2$ denotes the area element of the hyperbolic metric on $\widehat{\mathbb{C}} - P(f)$. Now $\operatorname{diam}(T_n)$ is bounded in terms of λ, so $\operatorname{area}(T_n)$ is bounded by a constant times $\operatorname{diam}(T_n)^2$. Thus

$$\frac{1}{\|(f^{k(n)})'(x)\|} \leq C(\lambda) \operatorname{diam}(T_n).$$

Combining this the bound $d(x, y) \leq C(\lambda)/\|(f^{k(n)})'(x)\|$ just obtained, we conclude that $d(x, y) \leq C(\lambda) \operatorname{diam}(T_n)$ for any y in T_n.

∎

Lemma 10.21 *The map h_n extends to a univalent map defined on an annulus of definite modulus about Y_n.*

Proof. By the Collar Theorem 2.18, there is a collar $C(\gamma_n)$ of definite modulus with core curve γ_n, contained in $\hat{\mathbb{C}} - P(f)$. Since $C(\gamma_n)$ is disjoint from the postcritical set, any univalent branch of an inverse iterate of f which is defined on Y_n extends to $C(\gamma_n)$. The map h_n is such an inverse branch.

∎

Completion of the proof of Theorem 10.14 (Proper invariant line field). Consider a sequence of n in $\mathcal{SR}(f, \lambda)$ such that Theorem 10.13 holds along this sequence. In particular, $(A_n(Y_n), 0) \to (Y, 0)$ in the Carathéodory topology, where $A_n(z) = z/\operatorname{Eucl. diam}(B_n)$ and $\operatorname{Eucl. diam}(\cdot)$ denotes Euclidean diameter.

Let
$$k_n = h_n \circ A_n^{-1} : A_n(Y_n) \to T_n.$$
By Theorem 10.12,
$$1 \le \operatorname{Eucl. diam}(A_n(Y_n)) \le C(\lambda).$$

Since k_n has a univalent extension to an annulus of definite modulus about $A_n(Y_n)$, and $k_n(A_n(Y_n)) = T_n$, the Koebe principle implies
$$\frac{1}{C(\lambda)}|k_n'(0)| \le \operatorname{Eucl. diam}(T_n) \le C(\lambda)|k_n'(0)|.$$

The Euclidean and hyperbolic metrics are nearly proportional near x, and the hyperbolic diameter of T_n tends to zero, so $|k_n'(0)| \to 0$. Similarly, $|x - k_n(0)|/|k_n'(0)|$ is bounded in terms of the ratio
$$\frac{d(x, T_n) + \operatorname{diam}(T_n)}{\operatorname{diam}(T_n)}$$
(measured in the hyperbolic metric), and the latter is bounded in terms of λ by Lemma 10.20.

Thus Theorem 5.16 implies there is a further subsequence of n in $\mathcal{SR}(f, \lambda)$ such that
$$\nu_n = k_n^*(\mu)$$
converges to a univalent line field ν on $(Y, 0) = \lim(A_n(Y), 0)$. But h_n is an inverse branch of an iterate of f, so f-invariance of μ implies $\nu_n = (A_n^{-1})^*\mu$. Since μ is f^n-invariant, the limit ν is g-invariant by Theorem 5.14.

∎

Proof of Theorem 10.9 (Thick rigidity). Suppose

$$0 < \liminf_{\mathcal{SR}(f)^*} \ell(\gamma_n) < \infty.$$

Then $|\mathcal{SR}(f, \lambda)| = \infty$ for some λ. If f admits an invariant line field μ on its Julia set, then by Theorems 10.13 and 10.14 we obtain a proper map of degree two $g : X \to Y$ with $g'(0) = 0$, $0 \in X \cap Y$, and a g-invariant univalent line field ν on Y. The presence of a critical point in Y makes this impossible (Theorem 5.13).

Thus f itself admits no measurable invariant line field on its Julia set.

∎

Proof of Theorem 1.7 (Robust rigidity). If f is robust, then $L = \liminf \ell(\gamma_n) < \infty$. If $L = 0$ then f admits no invariant line field on its Julia set by Corollary 10.4. The case $L > 0$ is covered by Theorem 10.9.

∎

Chapter 11

Real quadratic polynomials

This chapter recapitulates and carries further Sullivan's *a priori* bounds for the postcritical set of an infinitely renormalizable *real* quadratic polynomial [Sul4, §3]. In particular, we show any infinitely renormalizable real quadratic polynomial is robust. The main point is that the order structure of the real line keeps the postcritical set from doubling back on itself.

From robustness we deduce the main corollaries stated in the introduction. We conclude with a generalization to polynomials of the form $z^{2n} + c$.

Remark on bounded type. An infinitely renormalizable quadratic polynomial with $\mathcal{SR}(f) = \{n_0 < n_1 < n_2 < \dots\}$ has *bounded combinatorics* if $\sup n_{i+1}/n_i < \infty$. For *real* quadratics with bounded combinatorics, Sullivan also develops estimates for the quadratic-like renormalizations $f^n : U_n \to V_n$. These bounds are more difficult and will not be needed here.

11.1   Intervals and gaps

Let $f(z) = z^2 + c$, $c \in \mathbb{R}$ be a real quadratic polynomial with connected Julia set.

Theorem 11.1 *The filled Julia set $K(f)$ meets the real axis in an interval $[-\beta, \beta]$ bounded by the β fixed point of f and its preimage.*

Proof. By symmetry with respect to $z \mapsto \bar{z}$ and $z \mapsto -z$, the filled Julia set meets the real axis in an interval symmetric about the origin. The zero ray lies along the positive real axis, so the positive endpoint of this interval is β.

∎

Theorem 11.2 *Every renormalization of a real quadratic polynomial f is simple.*

Proof. Suppose $f^n : U_n \to V_n$ is a renormalization which is hybrid equivalent to $g(z) = z^2 + c'$. One may choose the quasiconformal conjugacy between f^n and g to respect the symmetry $z \mapsto \bar{z}$, so c' is real and $K(g) \cap \mathbb{R} = [-\beta_g, \beta_g]$, where β_g is the β fixed point of g. It follows that $K_n \cap \mathbb{R}$ is bounded by the β fixed point of $f^n : K_n \to K_n$ and its preimage. If K_n meets another small Julia set $K_n(i)$, it does so in a single periodic point x. This x is an endpoint of the interval $K_n \cap \mathbb{R}$, so x is the β fixed point of K_n and the renormalization is simple.

∎

Definitions. For a bounded subset $E \subset \mathbb{R}$ we let $[E]$ denote the closed convex hull of E (the smallest closed interval containing E). For each $n \in \mathcal{R}(f)$, let

$$I_n(i) \;=\; [P_n(i)].$$

Since $I_n(i)$ lies in the filled Julia set $K_n(i)$, which meets the real axis in an interval, the interiors of the intervals $I_n(i)$ at a given level are disjoint. If f is infinitely renormalizable, the closed intervals are disjoint as well, since the postcritical set contains no periodic points (Theorem 8.1). For $i < n$, f maps $I_n(i)$ homeomorphically to $I_n(i+1)$, while $f(I_n(n)) \subset I_n(1)$.

The *critical interval* at level n is $I_n(n)$, also denoted I_n. This is the interval containing the critical point.

Lemma 11.3 *Let $L(s) \supset I_n(s)$ be an open interval not containing $I_n(i)$ for any $i < s$. Then for $t = 1, \ldots s$ there is an interval*

$L(t) \supset I(t)$ *and a branch of* f^{t-s} *such that* $f^{t-s} : L(s) \to L(t)$ *is a homeomorphism.*

This branch extends to a univalent map $f^{t-s} : B(s) \to \mathbb{C}$, *where* $B(s)$ *is the open Euclidean ball with diameter* $L(s)$.

Remark. We emphasize that $L(s)$ can *meet* $I_n(i)$ for some $i < s$. For example, if $L(s)$ is the interior of the convex hull of the intervals adjacent to $I_n(s)$, then it satisfies the hypotheses of the lemma.

Proof. We will construct open intervals $L(t) \supset I(t)$ for $t = 1, \ldots s$, such that

(i) f maps $L(t)$ homeomorphically to $L(t+1)$ (for $t < s$); and

(ii) $L(t)$ does not contain $I_n(i)$ for any $i < t$.

By assumption $L(s)$ satisfies (ii).

Now suppose $L(t)$ has been constructed and $t > 1$. The critical value c of $f(z) = z^2 + c$ is the most negative point in the postcritical set (since $f(\mathbb{R}) = [c, \infty)$). Therefore (ii) implies c does not lie in $L(t)$ (if it did, $I_n(1)$ would be contained in $L(t)$.) Thus f^{-1} has two branches on $L(t)$, one of which maps $L(t)$ to an interval $L(t-1)$ containing $I_n(t-1)$. Property (ii) for $L(t-1)$ follows from (ii) for $L(t)$ and the fact that $f(I_n(i)) \subset I_n(i+1)$ for any i. Thus we may continue the induction until $t = 1$.

The postcritical set of f lies in $\mathbb{R} \cup \{\infty\}$, so any inverse branch defined on $L(s)$ can be extended to a univalent map on the upper and lower halfplanes. In particular, we may obtain a univalent map $f^{s-t} : B(s) \to \mathbb{C}$ extending the branch $f^{s-t} : L(s) \to L(t)$.

∎

Theorem 11.4 (Sullivan) *Suppose* f^n *is renormalizable, and let* $I_n(j)$ *denote the interval closest to the critical interval* I_n. *Then there is a universal* $\lambda > 1$ *such that*

$$|[I_n(j) \cup I_n]| > \lambda |I_n|.$$

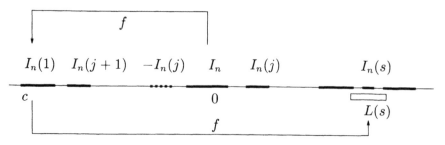

Figure 11.1. A definite gap to one side of the critical interval.

The proof is an application of the ubiquitous "shortest interval argument".

Proof of Theorem 11.4. Let $I_n(s)$ be the interval whose Euclidean length is the shortest among the intervals at level n. Let $L(s)$ be the open interval symmetric about $I_n(s)$ and of length $|L(s)| = 3|I_n(s)|$. Since $I_n(s)$ is the shortest interval, $L(s)$ does not contain any other intervals among the $I_n(i)$. Thus we may apply Lemma 11.3 to obtain a univalent branch of f^{1-s} defined on $B(s)$ and mapping $L(s)$ to $L(1) \supset I_n(1)$.

The Koebe distortion theorem implies $|L(1)| > \kappa |I_n(1)|$ for some universal $\kappa > 1$. Since $L(s)$ contains no $I_n(i)$ other than $I_n(s)$, the interval $L(1)$ only contains $I_n(1)$. Therefore the convex hull H of $I_n(1)$ and $I_n(j+1)$ satisfies

$$|H| > \kappa |I_n(1)|.$$

The interval $H' = f^{-1}(H)$ is the convex hull of $I_n(j)$ and $-I_n(j)$ (see Figure 11.1). Since f^{-1} is simply a square-root about c, we have

$$|H'| > \sqrt{\kappa} |I_n|.$$

The theorem follows, with $\lambda = (1 + \sqrt{\kappa})/2$.

■

For a more pictorial proof, see [Sul4, Figures 3 and 4].

Definition. A *gap* for $I_n(i)$ is an interval J, disjoint from $I_n(i)$, such that $[J \cup I_n(i)]$ is disjoint from $I_n(j)$ for every $j \neq i$. A *definite gap* is one satisfying $|J| > \alpha |I_n(i)|$ for a universal constant $\alpha > 0$.

In what follows, the implicit constant α for one theorem may depend on that from a preceding result, but since the bounds are derived in order, they are all universal.

Corollary 11.5 *If f^n is renormalizable, then every interval $I_n(i)$ has a definite gap to one side.*

Proof. The interval $-I_n(j)$ is disjoint from the postcritical set, so the critical interval I_n has a definite gap to one side (see Figure 11.1). Let $L(n) \supset I_n = I_n(n)$ be the interior of $[I_n(j) \cup -I_n(j)]$. Then we may apply Lemma 11.3 to construct intervals $L(i) \supset I_n(i)$ for $i = 1, \ldots n - 1$, and univalently branches of f^{i-n} defined on the ball with diameter $L(n)$ and sending $L(n)$ to $L(i)$. Since $\bigcup_{i=1}^{n} I_n(i)$ is forward invariant, the preimage of a gap is a gap. By the Koebe theorem, the image under f^{i-n} of the definite gap to one side of the critical interval gives a definite gap to one side of $I_n(i)$ for every i.

■

11.2 Real robustness

To obtain robustness, we need to have definite gaps on both sides of the critical interval. In general, however, such gaps are not present at every level of renormalization. For example, when one tunes $f_1(z) = z^2 - 1$ by an infinitely renormalizable real mapping $f_2(z) = z^2 + c$ with c close to -2, the resulting polynomial f has only a small gap between $I_2(1)$ and $I_2(2)$. This small gap results because the attracting basins of f_1 meet at the α fixed point. The construction cannot be iterated, however, because f_2 does not share this property. By this informal reasoning, one expects definite gaps on both sides to appear at least at every other level of renormalization.

More precisely, we have:

Theorem 11.6 *For every $n \in \mathcal{R}(f)$, either the critical interval I_n has a definite gap on both sides, or $j = n/2 \in \mathcal{R}(f)$ and I_j has a definite gap on both sides.*

Proof. As before, let $I_n(j)$ denote the interval closest to I_n. Consider the branch of f^{1-j} defined on $L(n) = [I_n(j) \cup -I_n(j)]$ that maps the critical interval I_n to $I_n(n - j + 1)$. Under this branch, the interval $I_n(j)$ either maps to a gap, or it maps to $I_n(1)$, the interval containing the critical value.

In the first case, by applying the Koebe theorem to f^{1-j}, we obtain definite gaps on both sides of $I_n(n - j + 1)$, since $I_n(j)$ and $-I_n(j)$ each have definite length compared to I_n. Similarly, the appropriate branch of f^{n-j} sends the definite gaps for $I_n(n - j + 1)$ to definite gaps on both sides of $I_n(1)$; pulling back by one more iterate of f, we obtain definite gaps on both sides of I_n. This establishes the theorem in the first case.

In the second case, we have that $I_n(1)$ and $I_n(n - j + 1)$ are adjacent intervals among $\bigcup I_n(i)$. Since f maps I_n to $I_n(1)$, it maps $I_n(j)$ to the interval adjacent to $I_n(1)$, and thus $j + 1 = n - j + 1$ and $j = n/2$.

Let $E_i = [I_n(i) \cup I_n(i + n/2)]$ for $i = 1, 2, \dots n/2$. It is not hard to check that these paired intervals are disjoint, so the sets E_i satisfy the hypotheses of Theorem 8.5, and therefore $f^{n/2}$ is renormalizable. Moreover $E_i = I_{n/2}(i)$.

To conclude, we must show that there is a definite gap on both sides of I_n or on both sides of $I_{n/2} = [I_n(n/2) \cup I_n]$. This is not hard to see: if $I_n(n/2)$ and I_n are too close together, then there must be definite space around $I_{n/2}$.

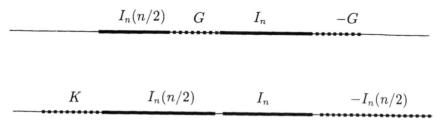

Figure 11.2. Definite gaps on both sides.

The argument is explained in Figure 11.2. Let G denote the interval between I_n and $I_n(n/2)$. By Theorem 11.4, $|G| + |I_n(n/2)|$ has definite size compared to $|I_n|$. If G is large enough, then G and $-G$ provide definite gaps on both sids of I_n and we are done. Otherwise $|G|$ is much smaller than $|I_n|$. In this case $|I_n(n/2)|$ must have definite size compared to $|I_n|$, so $|G|$ is also much smaller than $|I_n(n/2)|$. But then G is too small to serve as the definite gap for $I_n(n/2)$ guaranteed by Corollary 11.5. Thus there must be a definite gap K on the other side of $I_n(n/2)$. Since K has definite size compared to $I_n(n/2)$, and $I_n(n/2)$ has definite size compared to I_n, the intervals K and $-I_n(n/2)$ provide definite gaps on both sides of $I_{n/2} = [I_n(n/2) \cup I_n]$.

For the record, here are the constants. By Theorem 11.4,

$$|G| + |I_n(n/2)| = |G| + |I_n(j)| > \eta|I_n|,$$

for a universal constant $\eta = \lambda - 1 > 0$. By Corollary 11.5, there is a gap of length $\alpha|I_n(n/2)|$ to one side of $I_n(n/2)$ for a universal $\alpha > 0$.

Suppose $|G| \geq \alpha|I_n(n/2)|$. Then we may eliminate $|I_n(n/2)|$ from the inequality above to obtain

$$|G| > \frac{\eta}{1 + 1/\alpha}|I_n|,$$

so G and $-G$ provide definite gaps on both sides of I_n.

Now suppose $|G| < \alpha|I_n(n/2)|$. Then G is too small to serve as the definite gap for $I_n(n/2)$, so there is a gap K on the other side of $I_n(n/2)$ with $|K| > \alpha|I_n(n/2)|$. Moreover the length of $I_{n/2}$ is bounded above in terms of the length of $I_n(n/2)$:

$$\begin{aligned} |I_{n/2}| &= |I_n(n/2)| + |G| + |I_n| < (1 + 1/\eta)(|I_n(n/2)| + |G|) \\ &< (1 + 1/\eta)(1 + \alpha)|I_n(n/2)|. \end{aligned}$$

If we let $\zeta = (1 + 1/\eta)(1 + \alpha)$, then $|I_n(n/2)| > \zeta|I_{n/2}|$ and $|K| > \alpha\zeta|I_{n/2}|$, so K and $-I_n(n/2)$ provide definite gaps on both sides of $I_{n/2}$.

∎

Corollary 11.7 (Real robustness) *If f is an infinitely renormalizable real quadratic polynomial, then f is robust.*

Proof. By Theorem 11.2, all renormalizations of f are simple, so $|\mathcal{SR}(f)| = \infty$. Suppose f^n is renormalizable. The hyperbolic geodesic γ_n in $\widehat{\mathbb{C}} - P(f)$ is symmetric about the real axis and separates the critical interval I_n from the remaining intervals at level n. A definite gap on both sides of the critical interval gives an upper bound for the hyperbolic length $\ell(\gamma_n)$, by Theorem 2.3. (Alternatively, a definite gap gives an annulus of definite modulus in the homotopy class of γ_n; then $\ell(\gamma_n)$ is bounded above by the length of the core curve of this annulus.)

By Theorem 11.6, once $|\mathcal{SR}(f)| = \infty$ there are infinitely many n in $\mathcal{SR}(f)$ such that I_n has a definite gap on both sides. Therefore $\liminf_{\mathcal{SR}(f)} \ell(\gamma_n) < \infty$, so f is robust.

$\blacksquare$

11.3 Corollaries and generalizations

We may now deduce the two main corollaries stated in the introduction:

Corollary 1.8 *The Julia set of a real quadratic polynomial carries no invariant line field.*

Corollary 1.9 *Every component of the interior of the Mandelbrot set meeting the real axis is hyperbolic.*

Proof of Corollaries 1.8 and 1.9. Let $f(z) = z^2 + c$ be a real quadratic polynomial.

If f is infinitely renormalizable, then f is robust by Corollary 11.7, so by Theorem 1.7 (Robust rigidity) f carries no invariant line field on its Julia set. On the other hand, a map which is only finitely renormalizable carries no invariant line field by Corollary 8.7. This proves Corollary 1.8.

Now let U be a component of the interior of the Mandelbrot set meeting the real axis. By the preceding, f_c admits no invariant line field on its Julia set for c in $U \cap \mathbb{R}$. So by Theorem 4.9, U is hyperbolic.

$\blacksquare$

Although we have presented our rigidity argument for quadratic polynomials, many aspects generalize without effort to higher even exponents; for example one may establish:

Theorem 11.8 *If $f(z) = z^{2n} + c$, $n > 1$ is an infinitely renormalizable real polynomial, then f carries no invariant line field on its Julia set.*

However, at present one lacks a rigidity theory for *finitely* renormalizable mappings of higher degree (analogous to Theorem 1.6 for quadratic polynomials). It would also be natural to consider polynomials with several distinct critical points.

Appendix A

Orbifolds

This appendix provides a brief introduction to orbifolds. We begin with foundational material, including the uniformization theorem for Riemann orbifolds (Theorem A.4). Then we describe the orbifold $\mathcal{O}_f$ of a rational map f, and use it to discuss certain critically finite maps.

A.1 Smooth and complex orbifolds

An *orbifold* is a space which is locally modeled on the quotient of an open subset of $\mathbb{R}^n$ by the action of a finite group. For a general development see [Th1, §13]. We will need only the theory of smooth two dimensional orbifolds, and their complex analogues which generalize Riemann surfaces. Our definitions will take advantage of the simplifications possible in this case.

Definition. A *smooth n-dimensional orbifold* $\mathcal{O}$ is a Hausdorff topological space X together with an atlas $< U_\alpha, V_\alpha, \Gamma_\alpha, \phi_\alpha >$, where

1. $< U_\alpha >$ is an open covering of X, providing a base for the topology on X;

2. $< V_\alpha >$ is a collection of open subsets in $\mathbb{R}^n$;

3. Γ_α is a finite group of diffeomorphisms of V_α; and

4. $\phi_\alpha : V_\alpha \to U_\alpha$ is a continuous map whose fibers are the orbits of Γ_α.

This atlas is required to satisfy the following compatibility condition. Whenever $U_\alpha \subset U_\beta$, there exists an injective homomorphism $H_{\alpha\beta} : \Gamma_\alpha \to \Gamma_\beta$ and a smooth embedding $\phi_{\alpha\beta} : V_\alpha \to V_\beta$ such that:

1. for all $\gamma \in \Gamma_\alpha$ and $z \in V_\alpha$, we have $\phi_{\alpha\beta}(\gamma z) = H_{\alpha\beta}(\gamma)\phi_{\alpha\beta}(z)$; and

2. for all $z \in V_\alpha$, $\phi_\beta(\phi_{\alpha\beta}(z)) = \phi_\alpha(z)$.

The space X is called the *underlying space* of the orbifold $\mathcal{O}$.

A *complex n-orbifold* is defined by requiring that the charts V_α lie in $\mathbb{C}^n$, the transition functions $\phi_{\alpha\beta}$ are holomorphic and the groups Γ_α act biholomorphically.

Recall that a Riemann surface is a *connected* complex 1-manifold. Similar, we define a *Riemann orbifold* to be a connected one-dimensional complex orbifold.

Two atlases define the same orbifold structure if their union lies in a third atlas (satisfying the compatibility condition).

The sets V_α form the charts for the orbifold. Just as for a manifold, one studies local properties on an orbifold by passing to charts.

Specifying complex 1-orbifolds. A one-dimensional complex orbifold is conveniently specified by a pair (X, N) of a complex 1-manifold X and a *multiplicity map*

$$N : X \to \mathbb{N},$$

such that

$$\{x \in X \ : \ N(x) > 1\}$$

is discrete. To construct an orbifold from this data, consider the collection of all conformal isomorphisms $\psi_\alpha : \Delta \to U_\alpha \subset X$ such that $N(\psi_\alpha(z)) = 1$ for all z in Δ except possibly $z = 0$. Set $V_\alpha = \Delta$ and define $\phi_\alpha(z) = \psi_\alpha(z^n)$, where $n = N(\psi_\alpha(0))$. We then take Γ_α to be the group generated by $z \mapsto \exp(2\pi i/n)z$ acting on the disk, and the transition functions are determined in a straightforward way.

Every complex 1-orbifold $\mathcal{O}$ is specified in this manner. Indeed, a complex 1-orbifold $\mathcal{O}$ determines a complex manifold structure on the underlying surface X, because the quotients V_α/Γ_α carry natural complex structures. The orbifold also determines a map $N : X \to \mathbb{N}$ as follows: for $x \in U_\alpha$, choose y in V_α such that $\phi_\alpha(y) = x$, and set $N(x)$ equal to the cardinality of the stabilizer of y in Γ_α. It is easy to verify that $N(x)$ is well-defined, and that the original orbifold structure is equivalent to the one determined by the pair (X, N).

Similarly, every orientable 2-dimensional orbifold is specified by a pair (X, N) where X is a smooth surface and $N : X \to \mathbb{N}$ assumes values greater than one only on a discrete set. (Here *orientable* means there is an atlas such that the group actions and transition functions preserve orientation.)

A useful convention is to allow $N(x)$ to assume the value ∞ at a discrete set of points; these points are then omitted from the orbifold, so $(X, N) = (Y, N|Y)$ where $Y = \{x \ : \ N(x) < \infty\}$.

Definitions. Let $\mathcal{O} = (X, N)$ be a complex 1-orbifold. The *singular points* of $\mathcal{O}$ are those $x \in X$ with $N(x) > 1$. The *multiplicity* of a singular point x is $N(x)$. The *signature* of an orbifold is the list of values that N assumes at the singular points; a given value n is repeated as many times as the number of singular points of multiplicity n.

Traditional complex 1-manifolds will be regarded as orbifolds with $N(x) = 1$ everywhere.

A.2 Coverings and uniformization

In this section we treat holomorphic maps and covering maps between complex 1-orbifolds, and discuss the uniformization theorem.

Definitions. Let $f : X \to X'$ be a holomorphic map between complex 1-manifolds. The *local degree* $\deg(f, x)$ is equal to one more than the order to which f' vanishes at x. Thus $\deg(z^n, 0) = n$. (By convention the local degree is zero if f is locally constant at x.)

Let $\mathcal{O} = (X, N)$ and $\mathcal{O}' = (X', N')$ be complex 1-orbifolds. A *holomorphic map* from $\mathcal{O}$ to $\mathcal{O}'$ is a holomorphic map $f : X \to X'$ between the underlying complex manifolds such that:

$N(f(x))$ *divides* $\deg(f, x)N(x)$ *for each* x *in* X.

This condition is equivalent to the following local lifting property. Whenever $x' = f(x)$, there exist:

1. neighborhoods U_α and U'_α of x and x';

2. charts $\phi_\alpha : V_\alpha \to U_\alpha$ and $\phi'_\alpha : V'_\alpha \to U'_\alpha$; and

3. a holomorphic map $g : V_\alpha \to V'_\alpha$; such that

4. $f(\phi_\alpha(z)) = \phi'_\alpha(g(z))$ for all z in V_α.

A holomorphic map $f : \mathcal{O} \to \mathcal{O}'$ is a *covering map* if $N(f(x)) = \deg(f, x)N(x)$ for all x, and for every neighborhood U'_α of x', every component of $f^{-1}(U'_\alpha)$ maps surjectively to U'_α. Equivalently, the local lifting property can be verified using the same U'_α for every x with $f(x) = x'$, and $g : V_\alpha \to V'_\alpha$ can be chosen to be an embedding.

Now suppose $\mathcal{O} = (X, N)$ is connected. The *Euler characteristic* of $\mathcal{O}$ is given by

$$\chi(\mathcal{O}) \;=\; \chi(X) - \sum_X \left(1 - \frac{1}{N(x)} \right).$$

For an orbifold $\chi(\mathcal{O})$ is a rational number which may or may not be an integer. Intuitively, each point with $N(x) > 1$ contributes only $1/N(x)$ to the number of 0-cells in X. The Euler characteristic is $-\infty$ if $\pi_1(X)$ is infinitely generated or if X has infinitely many singular points.

The Euler characteristic satisfies $\chi(\mathcal{O}) = \deg(f)\chi(\mathcal{O}')$ for any finite degree covering map between orbifolds.

Theorem A.1 *Let Y be a Riemann surface, and let $\Gamma \subset \mathrm{Aut}(Y)$ be a group of automorphisms acting properly discontinuously. Then $X = Y/\Gamma$ carries a natural orbifold structure such that the projection $Y \to \mathcal{O} = (X, N)$ is a covering map.*

Proof. By proper discontinuity and removability of isolated singularities, $X = Y/\Gamma$ carries a Riemann surface structure such that the projection $\pi : Y \to X$ is holomorphic. Set $N(x)$ equal to the cardinality of the stabilizer of y, where $\pi(y) = x$. Since $N(x) = \deg(\pi, y)$, the projection to $\mathcal{O} = (X, N)$ is a covering map of orbifolds.

∎

Theorem A.2 (Uniformization) *Let $\mathcal{O}$ be a Riemann orbifold. Then exactly one of the following holds:*

1. $\mathcal{O}$ *is covered by* $\widehat{\mathbb{C}}$. *Equivalently,* $\mathcal{O}$ *is isomorphic to* $\widehat{\mathbb{C}}$ *or to a sphere with signature* (n, n), $(2, 2, n)$, $(2, 3, 3)$, $(2, 3, 4)$ *or* $(2, 3, 5)$, *where* $1 < n < \infty$.

2. $\mathcal{O}$ *is covered by* $\mathbb{C}$. *Equivalently,* $\mathcal{O}$ *is isomorphic to* $\mathbb{C}$, $\mathbb{C}^*$, *a complex torus, or a sphere with signature* (n, ∞), $(2, 2, \infty)$, $(2, 3, 6)$, $(2, 4, 4)$, $(3, 3, 3)$ *or* $(2, 2, 2, 2)$.

3. $\mathcal{O}$ *is covered by* $\mathbb{H}$. *Equivalently,* $\chi(\mathcal{O}) < 0$ *or* $\mathcal{O}$ *is isomorphic to an annulus of finite modulus, the unit disk, the punctured unit disk, a unit disk with signature* (n), $n > 1$, *or a unit disk with signature* $(2, 2)$.

4. $\mathcal{O}$ *is not covered by any Riemann surface. Equivalently,* $\mathcal{O}$ *is isomorphic to a sphere with signature* (n) *or* (n, m) *where* $1 < n < m < \infty$.

Proof. The universal covering orbifold $\mathcal{U}$ of $\mathcal{O} = (X, N)$ is constructed in [Th1, Theorem 13.2.4] by taking an inverse limit of fiber products of covering spaces of $\mathcal{O}$; it is unique up to isomorphism over $\mathcal{O}$. If $\mathcal{U}$ has no singular points, it is isomorphic to $\widehat{\mathbb{C}}$, $\mathbb{C}$, or $\mathbb{H}$ by the classification of simply-connected Riemann surfaces. On the other hand, if $\mathcal{U}$ does have singular points then $\mathcal{O}$ is not covered by any Riemann surface.

This shows exactly one of cases 1-4 holds. We now verify the equivalent formulations of these cases.

Singular case. The universal cover fails to be a Riemann surface if and only if $\mathcal{O}$ is a sphere with signature (n) or (n, m), $1 < n < m < \infty$. Indeed, if $\mathcal{O}$ is noncompact, then for any x in the singular locus a peripheral loop around x represents a nontrivial homology class in $H_1(X - \{x\}, \mathbb{R})$, so there is a finite regular covering orbifold of $\mathcal{O}$ where all the preimages of x are nonsingular. It follows that $\mathcal{U}$ has no singular points.

When $\mathcal{O}$ is compact, it has a finite covering which is a Riemann surface, apart from the examples enumerated in case 4. This observation goes back to Bundgaard, Nielsen and Fox; compare [Nam,

Theorem 1.2.15]. Thus the universal cover of $\mathcal{O}$ is also a Riemann surface.

It is easy to see directly that the examples in case 4 are not covered by any Riemann surface.

Elliptic case. Suppose the universal covering is given by $\pi : \widehat{\mathbb{C}} \to \mathcal{O}$. The group of deck transformations $\Gamma \subset \operatorname{Aut}(\widehat{\mathbb{C}})$ consists of those automorphisms such that $\pi \circ \gamma = \pi$; since $\widehat{\mathbb{C}}$ is compact, Γ has finite order. Up to conjugacy, Γ is a cyclic group, a dihedral group, or the group of symmetries of a regular tetrahedron, cube or dodecahedron. These groups give orbifolds with signatures (n, n), $(2, 2, n)$, $(2, 3, 3)$, $(2, 3, 4)$ and $(2, 3, 5)$ respectively. Conversely, any orbifold with these multiplicities is covered by $\widehat{\mathbb{C}}$, because any two triples of points on $\widehat{\mathbb{C}}$ are equivalent by a Möbius transformation.

Parabolic case. Suppose $\mathcal{O}$ is covered by $\mathbb{C}$. The group of deck transformations $\Gamma \subset \operatorname{Aut}(\mathbb{C})$ acts properly discontinuously on $\mathbb{C}$, so its subgroup Γ' of translations is normal and of finite index in Γ. Thus $\mathcal{O}$ admits a finite regular covering by $Z = \mathbb{C}/\Gamma'$, where Z is isomorphic to $\mathbb{C}$, $\mathbb{C}^*$ or a complex torus.

If $Z = \mathbb{C}$ then Γ is generated by a single rotation and $X = \mathbb{C}$ with at most one singular point.

If $Z = \mathbb{C}^*$, then $X = \mathbb{C}^*$ with no singular points or X is isomorphic to the quotient of Z by $z \mapsto 1/z$, which gives $\mathbb{C}$ with singular points of multiplicities $(2, 2)$.

If $Z = \mathbb{C}/\Gamma'$ is a torus, then typically the only non-translation automorphism of Z is conjugate to $z \mapsto -z$; this gives $X = \widehat{\mathbb{C}}$ with multiplicities $(2, 2, 2, 2)$. One exception arises when Γ' is isomorphic to $\mathbb{Z} \oplus \mathbb{Z}i$; then the quotient by $z \mapsto iz$ gives $X = \widehat{\mathbb{C}}$ with multiplicities $(2, 4, 4)$. Two others arise when Γ' is isomorphic to $\mathbb{Z} \oplus \mathbb{Z}\omega$, where ω is a primitive sixth root of unity; then the quotient by $z \mapsto \omega z$ gives multiplicities $(2, 3, 6)$, and the quotient by $z \mapsto \omega^2 z$ gives $(3, 3, 3)$.

The above discussion shows each of these orbifolds is indeed covered by $\mathbb{C}$, using the fact that any two triples of points on $\widehat{\mathbb{C}}$ are equivalent and any two-fold cover of $\widehat{\mathbb{C}}$ branched over four points is a torus.

Hyperbolic case. All Riemann orbifolds appearing in the preceding discussion have non-negative Euler characteristic. Thus all orbifolds with negative Euler characteristic, and all remaining Riemann orbifolds with non-positive Euler characteristic must be covered by $\mathbb{H}$. These remaining orbifolds are enumerated in case 3.

∎

Metrics on complex 1-orbifolds. A smooth Riemannian metric ρ on a complex 1-orbifold $\mathcal{O} = (X, N)$ is specified by a smooth metric ρ_α on each chart V_α, invariant under the action of Γ_α and compatible across charts. A smooth metric on $\mathcal{O}$ determines a singular metric on the complex manifold X. This metric has a singularity of the type

$$\rho = \frac{|dz|}{|z|^{1-1/n}}$$

near a singular point $z = 0$ of multiplicity n on X.

Definitions. A Riemann orbifold is *elliptic, parabolic* or *hyperbolic* if it is covered by $\widehat{\mathbb{C}}$, $\mathbb{C}$ or $\mathbb{H}$ respectively.

Such an orbifold inherits a spherical, Euclidean or hyperbolic metric from its universal cover. In the spherical and hyperbolic cases, this metric is uniquely determined by normalizing the curvature to ± 1. In the Euclidean case, the metric is well-defined up to a positive scalar multiple.

The Schwarz Lemma immediately generalizes to:

Theorem A.3 *Let $f : \mathcal{O} \to \mathcal{O}'$ be a holomorphic map between hyperbolic orbifolds. Then f does not increase the hyperbolic metric, and f is an infinitesimal isometry if and only if f is a covering map.*

A.3   The orbifold of a rational map

In §3 we introduced the hyperbolic metric on $\widehat{\mathbb{C}} - P(f)$ as a fundamental tool for studying the dynamics of a rational map f. For example, letting $Q(f) = f^{-1}(P(f))$, we observed that the restriction

$$f : (\widehat{\mathbb{C}} - Q(f)) \to (\widehat{\mathbb{C}} - P(f))$$

is a covering map, hence an isometry for the hyperbolic metric, while

$$(\widehat{\mathbb{C}} - Q(f)) \hookrightarrow (\widehat{\mathbb{C}} - P(f))$$

is holomorphic, hence a contraction. Putting these facts together, we saw that f expands the hyperbolic metric on $\widehat{\mathbb{C}} - P(f)$.

Orbifolds provide a refinement of this tool that is especially suited to rational maps with preperiodic critical points in the Julia set. We will construct orbifolds $\mathcal{O}_f$ and $\mathcal{Q}_f$ such that $f : \mathcal{Q}_f \to \mathcal{O}_f$ is a covering, and the inclusion $\mathcal{Q}_f \to \mathcal{O}_f$ is holomorphic. This method appears in [DH1, §III.7] and is developed systemically for critically finite maps in [Th2].

Definition. Let $f : \widehat{\mathbb{C}} \to \widehat{\mathbb{C}}$ be a rational map of degree $d > 1$ with postcritical set $P(f)$. The *orbifold* $\mathcal{O}_f = (X_f, N_f)$ of f is a complex 1-orbifold constructed as follows. First, X_f is obtained by deleting from $\widehat{\mathbb{C}}$ every non-isolated point of $P(f)$. Then, for each x in X_f, we define $N_f(x)$ to be the least common multiple of the local degrees $\deg(f^n, y)$ for all $n > 0$ and all $y \in \widehat{\mathbb{C}}$ such that $f^n(y) = x$. By convention $N_f(x) = \infty$ if these local degrees are unbounded; this happens if and only if x belongs to a periodic cycle containing a critical point.

Note that

$$\widehat{\mathbb{C}} - P(f) \subset X_f$$

and $N_f(x) = 1$ for every x outside $P(f)$, so this inclusion is a holomorphic map of orbifolds.

Example. Let $f(z) = z^2 + i$; then $X_f = \widehat{\mathbb{C}}$, $N_f(\infty) = \infty$, $N_f(i) = N_f(-1+i) = N_f(-i) = 2$, and $N_f = 1$ at all other points.

Now let $\mathcal{Q}_f = (Y_f, N_f')$ where $Y_f = f^{-1}(X_f)$ and $N_f'(y)$ is the integer $N_f(f(y))/\deg(f, y)$. It is easy to see

$$f : \mathcal{Q}_f \to \mathcal{O}_f$$

is a covering map, while the inclusion

$$\mathcal{Q}_f \to \mathcal{O}_f$$

is holomorphic as a map of orbifolds. This easily implies:

Theorem A.4 *Exactly one of the following holds:*

1. *The orbifold $\mathcal{O}_f$ is connected and parabolic, $f : \mathcal{O}_f \to \mathcal{O}_f$ is a covering map, and $\|f\| = C > 1$ in the Euclidean metric on $\mathcal{O}_f$.*

2. *Every component of $\mathcal{O}_f$ is hyperbolic, and $\|f'(x)\| \geq 1$ with respect to the hyperbolic metric on $\mathcal{O}_f$ when x and $f(x)$ both lie in $\mathcal{O}_f$.*

Proof. Suppose $\mathcal{O}_f$ is connected. Then

$$\chi(\mathcal{O}_f) \geq \chi(\mathcal{Q}_f) = \deg(f)\chi(\mathcal{O}_f),$$

so $\chi(\mathcal{O}_f) \leq 0$. If $\chi(\mathcal{O}_f) = 0$, then $\mathcal{O}_f$ is parabolic and the equality must hold above, which implies $\mathcal{Q}_f = \mathcal{O}_f$ and $f : \mathcal{O}_f \to \mathcal{O}_f$ is a covering map. The map f lifts to $z \mapsto \alpha z$ on the universal cover $\mathbb{C}$ of $\mathcal{O}_f$, where $C = |\alpha| > 1$ because the degree of f is greater than one. Thus f expands the Euclidean metric on $\mathcal{O}_f$ by the constant C.

If $\chi(\mathcal{O}_f) < 0$ or $\mathcal{O}_f$ is disconnected, then every component of $\mathcal{O}_f$ is hyperbolic and $\|f'\| \geq 1$ by the Schwarz Lemma.

∎

Rational maps with parabolic orbifolds are classified in [Th2] and [DH3]. A rich class of such examples are associated to the torus endomorphisms studied in §3.5, by the following result:

Theorem A.5 *If $\mathcal{O}_f$ is a sphere with signature $(2, 2, 2, 2)$, then f is double covered by a torus endomorphism.*

Proof. The orbifold $\mathcal{O}_f$ has a *canonical* two-fold covering space T which is a complex torus. (The torus T is just the two-fold branched covering of $\hat{\mathbb{C}}$ branched along $P(f)$. It can also be defined as the quotient of the universal cover $\mathbb{C}$ of $\mathcal{O}_f$ by those deck transformations which act as translations.)

Since $f : \mathcal{O}_f \to \mathcal{O}_f$ is a covering map, it lifts to an endomorphism of T.

∎

Example. Let $f(z) = (z^2 + 1 + \sqrt{2})/(z^2 - 1 - \sqrt{2})$. Then $\mathcal{O}_f$ is the sphere with singular points of multiplicity two at 1, -1, $1 + \sqrt{2}$ and $-1 - \sqrt{2}$. The map f is covered by the endomorphism $z \mapsto 2iz$ on the torus $T = \mathbb{C}/(\mathbb{Z} \oplus \mathbb{Z}\sqrt{2}i)$.

This example shows f need not be covered by an *integral* torus endomorphism.

Theorem A.6 *Let f be a critically finite rational map. Then each periodic cycle of f is repelling or superattracting. If f has no super-attracting cycles then $J(f) = \widehat{\mathbb{C}}$.*

Proof. The hypotheses imply $\mathcal{O}_f$ is a connected orbifold of finite Euler characteristic, uniformized by the Euclidean or hyperbolic plane. If $\mathcal{O}_f$ is parabolic, then $\|f'\| = C > 1$ with respect to the Euclidean metric on $\mathcal{O}_f$. If $\mathcal{O}_f$ is hyperbolic, then

$$\chi(\mathcal{Q}_f) = \deg(f)\chi(\mathcal{O}_f) < \chi(\mathcal{O}_f) < 0,$$

so the inclusion $\mathcal{Q}_f \to \mathcal{O}_f$ is not a covering map; by the Schwarz Lemma, we have $\|f'\| > 1$ pointwise with respect to the hyperbolic metric on $\mathcal{O}_f$.

In either case there is a Riemannian metric on $\mathcal{O}_f$ which is expanded by f, so each periodic cycle in $\mathcal{O}_f$ is repelling. Any periodic cycle which does not lie in $\mathcal{O}_f$ must contain a critical point, so it is superattracting.

In the absence of superattracting cycles, $\mathcal{O}_f$ is a compact orbifold (whose underlying space is the sphere), so we have uniform expansion in the orbifold metric. Thus the iterates of f cannot form a normal family at any point and the Julia set of f is $\widehat{\mathbb{C}}$.

∎

Compare [Th2, §13].
The preceding two theorems were used in §3.

Appendix B

A closing lemma for rational maps

In this appendix we develop the idea of a *quotient* of a rational map. A quotient is obtained by collapsing pieces of the postcritical set to single points, to yield a simpler branched covering of the sphere. We will show that when the postcritical set of the quotient is finite, the quotient branched covering can also be realized as a rational map.

Another way to look at our main result is that it provides a *closing lemma* for rational maps — that is, a method for finding a new dynamical system where a recurrent critical orbit becomes periodic. Our result will *not* guarantee that the closed dynamical system is near the original one, however.

For quadratic polynomials, there is a close relationship between quotients and renormalization. For example, we will associate to any infinitely renormalizable quadratic polynomial f an infinite sequence g_n of critically finite quotients. Conjecturally, these quotients g_n converge to the original map f. This conjecture implies the density of hyperbolic dynamics in the quadratic family.

B.1 Quotients of branched coverings

Definitions. To discuss the closing lemma, we introduce a category whose objects are branched coverings of the sphere and whose morphisms are combinatorial quotient maps.

Spheres with marked points. Let (S^2, A) denote the sphere with a distinguished closed subset A. A *quotient map*

$$\phi : (S^2, A) \rightarrow (S^2, B)$$

is a continuous, degree one map such that $\phi(A) = B$, $\phi^{-1}(b)$ is connected for all b in B, and $\phi^{-1}(x)$ is a single point for all x in $S^2 - B$. This means A is contained in a set obtained by "blowing up" points of B to (nonseparating) continua.

Two quotient maps ϕ_0 and ϕ_1 are *homotopic* if there is a continuous family ϕ_t of quotient maps connecting them, such that $\phi_t|A = \phi_0|A = \phi_1|A$.

A *combinatorial quotient map* is a homotopy class of quotient mappings. We obtain the category of *spheres with marked points* by taking the pairs (S^2, A) as objects and combinatorial quotient maps as morphisms.

It is well-known that if $|A| = |B| < \infty$, then any combinatorial quotient map $\phi : (S^2, A) \rightarrow (S^2, B)$ is represented by a homeomorphism, so these two objects are isomorphic. Similarly, the group of automorphisms of (S^2, A) is equal to the mapping class group of $S^2 - A$ if $|A|$ is finite.

We will use ϕ to denote both a combinatorial quotient map and a typical representative of its homotopy class, so long as the discussion is independent of the choice of representative.

Branched coverings. Let $f : S^2 \rightarrow S^2$ be a branched covering of the sphere; this means that f is a smooth map whose behavior is locally modeled on that of $z \mapsto z^d$ for some $d \geq 1$. The integer d is the *local degree* of f at x, denoted $\deg(f, x)$.

A rational map on the Riemann sphere is a special case of a branched covering.

The set of points with local degree greater than one form the *critical points* $C(f)$. The *postcritical set* $P(f)$ is closure of the forward orbits of the critical points:

$$P(f) = \overline{\bigcup_{c \in C(f), \, n > 0} f^n(c)}.$$

Note that the critical points need not belong to the postcritical set. We will be concerned exclusively with branched coverings f of degree two or more, in which case $|P(f)| \geq 2$.

A branched covering is *critically finite* if the $|P(f)| < \infty$.

Quotients and equivalence of branched coverings. Let f and g be two branched coverings on the sphere, and let ϕ be a combinatorial quotient map

$$\phi : (S^2, P(f)) \to (S^2, P(g)).$$

Suppose there are maps ϕ_0 and ϕ_1 in the homotopy class ϕ making the diagram

$$
\begin{array}{ccc}
S^2 & \xrightarrow{\phi_1} & S^2 \\
{\scriptstyle f}\downarrow & & \downarrow{\scriptstyle g} \\
S^2 & \xrightarrow{\phi_0} & S^2
\end{array}
$$

commutative. In this case we say that g is a *quotient* of f, and write

$$\phi : f \to g.$$

We can then form the category of *branched coverings of the sphere* whose objects are branched coverings f and whose morphisms are combinatorial quotient maps as above.

The map g is a *proper* quotient of f if $|\phi^{-1}(x) \cap P(f)| > 1$ for some x in $P(g)$. This means at least two points of $P(f)$ are collapsed to form a single point of $P(g)$.

Combinatorial equivalence. Thurston defined two critically finite branched coverings f and g to be *combinatorially equivalent* if there are homeomorphisms $\phi_i : (S^2, P(f)) \to (S^2, P(g))$, $i = 0, 1$, such that the diagram above commutes, and ϕ_0 is isotopic to ϕ_1 rel $(P(f), P(g))$ [Th2], [DH3]. Alternatively, f and g are combinatorially equivalent if after deforming f by isotopy rel $P(f)$, it becomes topologically conjugate to g. Using techniques from the mapping class groups of surfaces, it can be shown that Thurston's notion of combinatorial equivalence agrees with isomorphism in the category of branched coverings introduced above.

Thus we will use the term *combinatorial equivalence* to denote isomorphism in this category.

We can now formulate our main result.

Theorem B.1 (Rational quotients) *Let g be a critically finite quotient of a rational map f. Then g is combinatorially equivalent to a rational map.*

B.2   Critically finite rational maps

The proof of Theorem B.1 is an application of Thurston's characterization of critically finite rational maps. To state this characterization, we introduce the orbifold associated to a branched covering
and the eigenvalue of an invariant curve system.

The orbifold of a branched cover.   Let $f : S^2 \to S^2$ be a
critically finite branched covering. As we did for rational maps in
§A.3, we will define a *smooth* orbifold canonically associated to f.

For each $x \in S^2$ define $N_f(x)$ (which may be ∞) as the least
common multiple of the local degrees $\deg(f^n, y)$ for all $n > 0$ and all
y in S^2 such that $f^n(y) = x$. (Note that $N_f(x) = 1$ if x is not in the
postcritical set $P(f)$.) Then $\mathcal{O}_f = (S^2, N_f)$ is *orbifold* of f.

The eigenvalue of a curve system.   A simple closed curve γ on
$S^2 - P(f)$ is *essential* if it does not bound a disk in $S^2 - P(f)$. A
curve is *peripheral* if it encloses a single point of $P(f)$. Two simple
curves γ and δ are *parallel* if they are isotopic in $S^2 - P(f)$.

A *curve system* $\Gamma = \{\gamma_i\}$ on $S^2 - P(f)$ is a finite nonempty
collection of disjoint simple closed curves, each essential and nonperipheral, and no two parallel. A curve system determines a *transition
matrix* $A(\Gamma) : \mathbb{R}^\Gamma \to \mathbb{R}^\Gamma$ by the formula

$$A_{\gamma\delta} \;=\; \sum_\alpha \frac{1}{\deg(f : \alpha \to \delta)}$$

where the sum is taken over components α of $f^{-1}(\delta)$ which are isotopic to γ.

Let $\lambda(\Gamma) \geq 0$ denote the spectral radius of $A(\gamma)$. Since $A(\Gamma) \geq 0$,
the Perron-Frobenius theorem guarantees that $\lambda(\Gamma)$ is an eigenvalue
for $A(\Gamma)$ with a non-negative eigenvector [Gant, §XIII].

A curve system is f-*invariant* if for each γ in Γ, each component
α of $f^{-1}(\gamma)$ is either inessential, peripheral or parallel to a curve in
Γ.

Theorem B.2 (Thurston) *Let* $f : S^2 \to S^2$ *be a critically finite
branched covering. Then* f *is combinatorially equivalent to a rational
map* g *if and only if:*

(Torus case) $\mathcal{O}_f$ *has signature* $(2, 2, 2, 2)$ *and* g *is double covered
by a torus endomorphism; or*

(General case) $\mathcal{O}_f$ *does not have signature* $(2, 2, 2, 2)$ *and*

$$\lambda(\Gamma) < 1$$

for every f-invariant curve system Γ on $S^2 - P(f)$.

In the second case, g is unique up to conformal conjugation.

For a proof, see [Th2], [DH3].

Remark. In the torus case, f lifts to a covering map $\tilde{f} : X \to X$ where X is a torus double covering $\mathcal{O}_f$. (Compare Theorem A.5.) Then f is equivalent to a rational map if and only if the induced linear map

$$\tilde{f}_* : H_1(X, \mathbb{R}) \to H_1(X, \mathbb{R})$$

is conformal with respect some metric on $H_1(X, \mathbb{R}) \cong \mathbb{R}^2$. Equivalently, if we represent $\tilde{f}_*$ by a 2×2 integer matrix $\begin{pmatrix} a & b \\ c & d \end{pmatrix}$, then f is combinatorially a rational map if and only if the corresponding Möbius transformation $z \mapsto (az + b)/(cz + d)$ has a fixed point in the upper halfplane. Thus we have a combinatorial criterion for rationality in the torus case as well as in the general case.

In the torus case, $\mathbb{H}$ is the Teichmüller space of conformal structures on $\mathcal{O}_f$, pullback of structures by f determines a map of Teichmüller space to itself, and f is combinatorially rational if and only if this map has a fixed point. The general case is analyzed by a similar method, using the Teichmüller space of the sphere with $|P(f)|$ marked points.

To show that a critically finite quotient g of a rational map f is itself rational (Theorem B.1), the basic idea is to check that g inherits the property $\lambda(\Gamma) < 1$ from f and apply Thurston's criterion. There are two important details to take care of:

1. The rational map f might admit an invariant curve system with $\lambda(\Gamma) = 1$.

2. The orbifold $\mathcal{O}_g$ of the branched covering g might have signature $(2, 2, 2, 2)$.

These details are addressed in the next two sections to complete the proof of Theorem B.1 (Rational quotients).

B.3   Siegel disks, Herman rings and curve systems

In this section we consider a rational map f whose postcritical set may be infinite, and classify all the examples which admit a curve system Γ with $\lambda(\Gamma) = 1$.

We *will not* require that Γ is f-invariant. Although Thurston's theorem makes reference only to invariant curve systems, it is technically convenient to forgo this assumption.

Theorem B.3 *Let $f(z)$ be a rational map, and let Γ be a curve system on $\widehat{\mathbb{C}} - P(f)$. Then $\lambda(\Gamma) \le 1$.*

Theorem B.4 (Classification of $\lambda(\Gamma) = 1$) *Let Γ be a curve system for a rational map f. If $\lambda(\Gamma) = 1$, then either:*

1. *f is critically finite, $\mathcal{O}_f$ is the $(2, 2, 2, 2)$ orbifold and f is double covered by an integral torus endomorphism; or*

2. *$|P(f)| = \infty$, and Γ includes the essential curves in a finite system of annuli permuted by f. These annuli lie in Siegel disks or Herman rings for f, and each annulus is a connected component of $\widehat{\mathbb{C}} - P(f)$.*

Definitions. Disjoint annuli $A_1, \ldots A_n$ on the sphere are *nested* if there are two points which are separated by every A_i. The *join* of a nested sequence, denoted $\mathrm{join}(A_1, \ldots, A_n)$, is the smallest annulus containing every A_i as an essential subannulus. Its boundary consists of a one component from the boundary of the "innermost" annulus and another from the "outermost" annulus.

Theorem B.5 *If $B = \mathrm{join}(A_1, \ldots, A_n)$ is the join of a set of nested annuli of finite moduli, then*

$$\mathrm{mod}(B) \ge \sum \mathrm{mod}(A_i).$$

Equality holds if and only if in a conformal coordinate where $B = \{z \,:\, 1 < |z| < R\}$, each A_i has the form

$$A_i = \{z \,:\, r_i < |z| < s_i\},$$

and the A_i fill all of B except for a finite number of circles.

This proposition is a sharp form of superadditivity of the modulus [LV, §I.6.6].

Now let Γ be a curve system with transition matrix $A_{\gamma\delta}$. We say Γ is *irreducible* if for any (γ, δ) there is an $n > 0$ such that $A^n_{\gamma\delta} > 0$. The Perron-Frobenius theory easily implies [Gant, §XIII.4]:

Theorem B.6 *Any curve system with* $\lambda(\Gamma) > 0$ *contains an irreducible curve system* Γ' *with* $\lambda(\Gamma) = \lambda(\Gamma')$.

Thus in the proofs of Theorems B.3 and B.4 we will assume that Γ is irreducible.[1]

Both proofs involve the study of systems of disjoint simple annuli C_γ representing the isotopy classes Γ on $\widehat{\mathbb{C}} - P(f)$. From any system of annuli C_γ, $\gamma \in \Gamma$, we can construct a new system of *pullback annuli* C'_γ by setting

$$C'_\gamma = \mathrm{join}(D_1, \dots, D_n),$$

where D_i enumerates the set of components of $f^{-1}(\bigcup_{\delta \in \Gamma} C_\delta)$ which are isotopic to γ in $\widehat{\mathbb{C}} - P(f)$. (This set is nonempty by irreducibility). We denote this pullback operation by

$$C' = f^*C.$$

Theorem B.7 *The moduli of the pullback annuli satisfy*

$$\mathrm{mod}(C'_\gamma) \geq \sum_\gamma A_{\gamma\delta} \, \mathrm{mod}(C_\delta).$$

Proof. If an annulus A' covers an annulus A with degree d, then $\mathrm{mod}(A') = \mathrm{mod}(A)/d$; the proposition follows from this fact, superadditivity of the modulus and the definition of the transition matrix.

∎

---

[1] The property of f-invariance may be lost in passing to Γ', which is one reason we do not require invariance.

Proof of Theorem B.3. The proof follows the same lines as part of Thurston's result (Theorem B.2). There are constants M_γ such that

$$\text{mod}(C_\gamma) \leq M_\gamma$$

for any annulus system representing Γ; for example, if X is the component of $\widehat{\mathbb{C}} - P(f)$ containing γ, we may choose M_γ equal to the modulus of the covering space of X determined by the cyclic subgroup $\langle \gamma \rangle \subset \pi_1(X)$. Starting with any annulus system C^0, define inductively $C^{n+1} = f^*C^n$. Since the modulus of C_γ^n is bounded above, the iterates of A applied to the positive vector vector $[\text{mod}(C_\gamma^0)]$ are bounded as well, so $\lambda(\Gamma) \leq 1$.

∎

Now assume $\lambda(\Gamma) = 1$. To analyze this case, it is useful to choose the annuli C_γ as large as possible, using:

Theorem B.8 (Strebel) *Let Γ be a nonempty set of disjoint simple geodesics on a (connected) hyperbolic Riemann surface X with assigned weights $m_\gamma > 0$. Then there exists a unique collection of disjoint open annuli C_γ representing the isotopy classes Γ and maximizing $\sum \text{mod}(C_\gamma)$, subject to the condition that the moduli $[\text{mod}(C_\gamma)]$ are proportional to $[m_\gamma]$.*

From uniqueness one may easily show:

Corollary B.9 *If C_γ' is another system of disjoint annuli representing Γ, such that $\text{mod}(C_\gamma') \geq \text{mod}(C_\gamma)$ for all γ, then $C_\gamma' = C_\gamma$.*

See [Str, Theorems 20.6 and 21.7].

Now let $m_\gamma > 0$ be a positive solution to the eigenvalue equation $Am = m$; such a solution exists by irreducibility. Applying Strebel's theorem, we will construct a canonical system of annuli C_γ representing the curves Γ, and with $[\text{mod}(C_\gamma)]$ proportional to $[m_\gamma]$.

More precisely, we define C_γ as the unique system of annuli such that

(a) $\sum \text{mod}(C_\gamma)$ is maximized, subject to the condition:

(b) $\text{mod}(C_\gamma)/\text{mod}(C_\delta) = m_\gamma/m_\delta$ whenever γ and δ lie in the same component of $\widehat{\mathbb{C}} - P(f)$.

Although we have only applied Strebel's result component by component, we have:

Theorem B.10 *Assuming Γ is irreducible, there is a constant $c > 0$ such that* $\mathrm{mod}(C_\gamma) = cm_\gamma$ *for all* γ.

Proof. Write $\mathrm{mod}(C_\gamma) = cm_\gamma + v_\gamma$, where $c > 0$, $v_\gamma \geq 0$ and $v_\alpha = 0$ for some particular curve α lying in a component X of $\widehat{\mathbb{C}} - P(f)$. Then by construction, $\mathrm{mod}(C_\gamma) = cm_\gamma$ for every γ lying in X.

We will show $v = 0$. If not, we can choose n such that $(A^n v)_\alpha > 0$ by irreducibility. Then $C' = (f^n)^*(C)$ satisfies $\mathrm{mod}(C'_\alpha) > \mathrm{mod}(C_\alpha)$ and $\mathrm{mod}(C'_\gamma) \geq \mathrm{mod}(C_\gamma)$ for all γ lying in X, contradicting Corollary B.9. ∎

Corollary B.11 *The maximal annuli are invariant under pullback: if $C' = f^*C$, then $C'_\gamma = C_\gamma$ for all $\gamma \in \Gamma$.*

Proof. Since $Am = m$, the moduli of the pullback annuli satisfy $\mathrm{mod}(C'_\gamma) \geq \mathrm{mod}(C_\gamma)$; thus the two systems of annuli are equal by the uniqueness part of Strebel's theorem (Corollary B.9). ∎

Definition. Let $D = f^{-1}(\bigcup_\Gamma C_\gamma)$. An *amalgam* A is an annulus in $\widehat{\mathbb{C}}$ such that $A = \mathrm{join}(D_1, \ldots, D_n)$, and $\mathrm{mod}(A) = \sum \mathrm{mod}(D_i)$, for some collection $D_1, \ldots, D_n$ of nested components of D. It follows that adjacent D_i are separated by real analytic circles as in Theorem B.5. Note that the D_i's *may* represent distinct homotopy classes on $\widehat{\mathbb{C}} - P(f)$.

Since $\mathrm{mod}(C_\gamma) = \sum A_{\gamma\delta} \mathrm{mod}(C_\delta) = \sum \mathrm{mod}(D_i)$, where the last sum is over the components of D homotopic to γ, we have:

Theorem B.12 *Every annulus C_γ is an amalgam.*

Theorem B.13 *Let A be an amalgam which does not meet the critical points $C(f)$. Then $f(A)$ is an amalgam and the map $A \to f(A)$ is a covering map.*

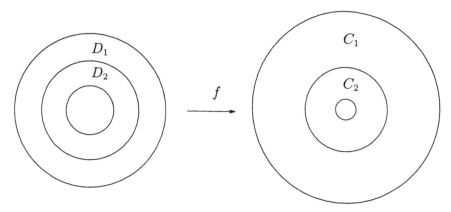

Figure B.1.   An unbranched image of an amalgam is an amalgam.

Proof. Let A be an amalgam of $D_1, \ldots D_k$. Then f maps each D_i by a covering map to C_i, one of the original annuli C_γ. If there are no critical points on the circles separating adjacent D_i's, then the C_i's are disjoint and nested, so A maps to $f(A)$ by a covering map (see Figure B.1). Then $\mathrm{mod}(f(A)) = \sum \mathrm{mod}(C_i)$, and since each C_i is an amalgam, so is $f(A)$.

∎

Theorem B.14 *Let A be an amalgam which meets the critical points $C(f)$, and let B_1, B_2 denote the components of ∂A. Then there are amalgams A_1, A_2 in A, disjoint from $C(f)$, such that $\partial f(A_i) = f(B_i) \sqcup I_i$ where I_1, I_2 are real analytic intervals, and the endpoints of I_i lie in the postcritical set $P(f)$.*

Proof. Let $A = \mathrm{join}(D_1, \ldots D_n)$; since each D_i maps by a covering map, the critical points of f must be contained in the circles separating adjacent D_i's. Let A_i be the maximal amalgam in A disjoint from $C(f)$ and containing B_i as one of its boundary components. (This A_i is the join of one or more adjacent D_i's lying near one end of A.)

   Consider the circle S_i forming the other boundary component of A_i; it necessarily meets the critical points of f. The circle S_i is

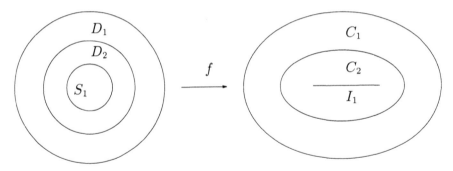

Figure B.2. A branched image of an amalgam, bounded by an interval.

a component of $f^{-1}(\widehat{\mathbb{C}} - \bigcup C_\gamma)$, so it is a branched cover of $f(S_i)$. Therefore $f(S_i) = I_i$, an interval whose endpoints are critical values (see Figure B.2). By the preceding result, $A_i \to f(A_i)$ is a finite covering map, hence proper, so

$$\partial f(A_i) = f(\partial A_i) = f(B_i) \sqcup f(S_i).$$

■

Proof of Theorem B.4 (Classification of $\lambda(\Gamma) = 1$). Suppose there exists an annulus C_γ whose iterates $A_n = f^n(C_\gamma)$ are disjoint from the critical points of f for all $n \geq 0$. Then each A_n is an amalgam, f carries A_n to A_{n+1} by a covering map, and $\mathrm{mod}(A_{n+1}) \geq \mathrm{mod}(A_n)$. There are only finitely many possible amalgams, so $A_n = A_m$ for some $n > m > 0$ and f^{n-m} maps A_n to itself by degree one. Since A_n is the join of annuli from among the C_γ, some annulus C_α is mapped to itself by degree one by an iterate of f. Therefore C_α is contained in a Siegel disk or Herman ring.

For any Herman ring or Siegel disk U, $\partial U \subset P(f)$ and $P(f) \cap U$ is invariant under rotation of U. Thus the assertions of case 2 of the Theorem are easily verified.

Now assume the forward orbit of any annulus C_γ eventually encounters a critical point. Then there is an annulus C_α with one boundary component equal to an interval I. Consider the least $n \geq 0$

such that $A = f^n(C_\alpha)$ contains a critical point of f. Then A is an amalgam and one boundary component B_1 of A is an interval. By Theorem B.14, there is an amalgam $A_1 \subset A - C(f)$ such that $\partial(f(A_1)) = f(B_1) \cup I_1 = J_1 \cup I_1$ is a pair of intervals. It follows that $\widehat{\mathbb{C}} = A' \cup I_1 \cup J_1$, where A' is the amalgam $f(A_1)$.

To complete the proof, one may check that the postcritical set of f is equal to the set of endpoints of I_1 and J_1, and $\mathcal{O}_f$ is the $(2, 2, 2, 2)$ orbifold. Then f is covered by a map $F(z) = \alpha z$ on a complex torus. Since f admits an invariant curve system, α is an integer.

(Alternatively, one can consider the canonical quadratic differential ϕ which comes along with Strebel's result [Str, Theorem 21.7]. It can be shown that ϕ extends to a meromorphic differential on $\widehat{\mathbb{C}}$ satisfying $f^*\phi = \deg(f)\phi$, which also implies case 1 of the Theorem by Lemma 3.16.)

∎

The proof also yields:

Corollary B.15 *Let f be a rational map, and let $\gamma \subset \widehat{\mathbb{C}} - P(f)$ be an essential nonperipheral simple closed curve.*

Suppose f maps δ to γ by degree one, where δ is a component of $f^{-1}(\gamma)$ isotopic to γ. Then γ lies in an annular component of $\widehat{\mathbb{C}} - P(f)$, contained in a Siegel disk or Herman ring for f.

∎

B.4   Rational quotients

Definition. Let $\phi : f \to g$ be a quotient map. A point $x \in P(g)$ is *blown-up* by ϕ if $|\phi^{-1}(x) \cap P(f)| > 1$. These are the postcritical points which are properly modified by ϕ to obtain f.

Theorem B.16 *For any quotient map $\phi : f \to g$ between branched coverings of the sphere, the set $B \subset P(g)$ of blown-up postcritical points is forward invariant ($g(B) \subset B$).*

Proof. We will show the points which are not blown-up are backward invariant. Suppose $x \in P(g)$ is not blown-up, and $g(y) = x$, where y is also in $P(g)$. Choose a disk U meeting $P(g)$ only in x, and let V be the component of $g^{-1}(U)$ such that $V \cap P(g) = \{y\}$. Then $\phi^{-1}(U)$ meets $P(f)$ in at most a single point, so the same is true of each component of $f^{-1}\phi^{-1}(U)$, one of which is $\phi^{-1}(V)$. Therefore y is not blown-up by ϕ.

∎

Theorem B.17 *Let g be a critically finite branched covering which is a quotient of a rational map f, and let $B \subset P(g)$ denote the blown-up postcritical points. Then every periodic cycle of B contains a critical point.*

Note that when B is nonempty it always has periodic cycles, since $g(B) \subset B$.

Proof. If $|P(g)| = 2$, then g is combinatorially equivalent to z^n and the conclusion is immediate.

Now assume $|P(g)| > 2$ and B contains a periodic cycle without a critical point; we will deduce a contradiction.

Replacing f and g by appropriate iterates, we may assume B contains a fixed point x which is not a critical point. Let γ be a peripheral curve around x; then $g^{-1}(\gamma)$ contains a peripheral curve δ around x mapping to γ by degree one. Applying Corollary B.15 to $\gamma' = \phi^{-1}(\gamma)$, we find that γ' lies in an annular component X of $\widehat{\mathbb{C}} - P(f)$. But ϕ maps X homeomorphically to $\widehat{\mathbb{C}} - P(g)$, contradicting the assumption that $|P(g)| > 2$.

∎

Corollary B.18 *If O_g is the Euclidean $(2, 2, 2, 2)$ orbifold, then the branched covering g is not a proper quotient of any rational map f.*

Proof. The map g has no periodic critical points, so no points in $P(g)$ are blown-up.

∎

Proof of Theorem B.1 (Rational quotients). Let $\phi : f \rightarrow g$ be a quotient map between a rational map f and a critically finite branched cover g. We may assume that g is a proper quotient of f.

If $|P(g)| \leq 3$ then g is rational (by Thurston's characterization, or more simply by uniqueness of the conformal structure on a sphere with three or fewer marked points).

If $|P(g)| \geq 4$, then $|P(f)| \geq 5$ and by the corollary above, the signature of $\mathcal{O}_g$ is not $(2, 2, 2, 2)$. So to prove g is rational, it suffices to show $\lambda(\Gamma) < 1$ for every g-invariant curve system.

Let Γ be a g-invariant curve system. Then $\Gamma' = \phi^{-1}(\Gamma)$ is a curve system for f and $\lambda(\Gamma') = \lambda(\Gamma)$.[2]

By Theorem B.4 and the fact that $|P(f)| > 4$, we see either $\lambda(\Gamma') < 1$ or Γ' includes a curve lying in an annular component X of $\widehat{\mathbb{C}} - P(f)$. But in the latter case X is homeomorphic to $\widehat{\mathbb{C}} - P(g)$, which is impossible (as before) because $|P(g)| > 2$.

■

B.5   Quotients and renormalization

To conclude, we will show that for quadratic polynomials there is a close relationship between quotients and renormalization.

Theorem B.19 *Let g be a critically finite quotient of a quadratic polynomial f. Then the Julia set of f is connected, and g is combinatorially equivalent to a unique quadratic polynomial $z^2 + c'$.*

If the quotient is proper, then the critical points of g are periodic and $z^2 + c'$ is hyperbolic.

Proof. Let $\phi : f \rightarrow g$ be a quotient map. By Theorem B.1, up to combinatorial equivalence we may replace g by a rational map. Since $f^{-1}(\infty) = \infty$, g leaves the point $\phi(\infty)$ totally invariant, and thus we may normalize by a Möbius transformation so that $g(z) = z^2 + c'$. Because g is critically finite, its Julia set is connected, and $\phi^{-1}(P(g) - \{\infty\})$ is a compact subset of $\mathbb{C}$ containing the forward

---

[2] The curve system Γ' may not be f-invariant, because ϕ^{-1} of a peripheral curve need not be peripheral.

orbit of the critical value $z = 0$ of f, so the Julia set of f is also connected.

If g is a proper quotient of f, the set $B \subset P(g)$ of blown-up postcritical points is nonempty; since $\phi^{-1}(\infty) = \infty$, we have $B \subset \mathbb{C}$. By Theorem B.16 and Theorem B.17, the set B contains a periodic critical point of g, which must be $z = 0$. Therefore both critical points of g are periodic and g is hyperbolic.

∎

Theorem B.20 (Renormalizable implies divisible) *Let $f(z) = z^2 + c$, and suppose f^n is simply renormalizable, with disjoint small postcritical sets $P_n(1), \ldots P_n(n)$. Then there exists a natural quotient map $\phi : f \to g$, where $g(z) = z^2 + c'$ is a quadratic polynomial with a superattracting cycle of period n.*

Proof. By Theorem 9.2, there is a system of n disjoint simple closed curves $\Gamma_n = \{\gamma_n(1), \ldots, \gamma_n(n)\}$ in $\widehat{\mathbb{C}} - P(f)$, with $\gamma_n(i)$ bounding a disk D_i such that $D_i \cap P(f) = P_n(i)$. Moreover $f^{-1}(\Gamma_n)$ contains n curves $\{\alpha_n(1), \ldots, \alpha_n(n)\}$ with $\alpha_n(i)$ isotopic to $\gamma_n(i)$ on $\widehat{\mathbb{C}} - P(f)$. Thus there is an isotopy

$$h_t : (\widehat{\mathbb{C}}, P(f)) \to (\widehat{\mathbb{C}}, P(f))$$

such that $h_0 = \mathrm{id}$ and $h_1(\alpha_n(i)) = \gamma_n(i)$.

Let $P(g) = \{p_1, \ldots, p_n, q\} \subset S^2$ be a set of $n+1$ distinct points, and let

$$\psi : (\widehat{\mathbb{C}}, P(f)) \to (S^2, P(g))$$

be a continuous map such that $\psi^{-1}(p_i) = D_i$, ψ is injective outside $\bigcup D_i$, and $\psi(\infty) = q$. Then ψ is a quotient map.

Let

$$\phi_t = \psi \circ h_t : (\widehat{\mathbb{C}}, P(f)) \to (S^2, P(g)).$$

We claim there is a branched covering g making the diagram

$$
\begin{array}{ccc}
(\widehat{\mathbb{C}}, P(f)) & \xrightarrow{\ \phi_1\ } & (S^2, P(g)) \\[4pt]
f \downarrow & & \downarrow g \\[4pt]
(\widehat{\mathbb{C}}, P(f)) & \xrightarrow{\ \phi_0\ } & (S^2, P(g))
\end{array}
$$

commute. Indeed, g can be defined by $g(x) = \psi \circ f \circ h_1^{-1} \circ \psi^{-1}(x)$. To see this definition is unambiguous, suppose $x = p_i$. Then $\psi^{-1}(x) = D_i$; $h_1^{-1}(D_i)$ is bounded by $\alpha_n(i)$, and maps to D_{i+1} under f; and ψ collapses D_{i+1} to p_{i+1}.

It is easy to verify that $P(g)$ is the postcritical set of g. Since ϕ_t provides a homotopy between ϕ_0 and ϕ_1, we have determined a quotient map $\phi : f \to g$. By the preceding result, up to combinatorial equivalence we may replace g by a quadratic polynomial $z^2 + c'$ with a superattracting cycle of period n.

∎

Corollary B.21 *If* $f(z) = z^2 + c$ *is an infinitely renormalizable quadratic polynomial, then* f *admits infinitely many distinct critically finite quotients* $g_n(z) = z^2 + c_n$.

Proof. By Theorem 8.4, f^n is simply renormalizable for infinitely many n, so by the preceding result there are critically finite quotients g_n with $|P(g_n)| \to \infty$.

∎

We can now formulate:

Conjecture B.22 *For any infinitely renormalizable* f, *the critically finite quotients* g_n *converge to* f.

Theorem B.23 *Conjecture B.22 implies the density of hyperbolic dynamics in the quadratic family* $z^2 + c$.

Proof. Suppose hyperbolic dynamics is not dense. Then, by Corollary 4.10, there is a quadratic polynomial $f(z) = z^2 + c$ which carries a measurable invariant line field on its Julia set. By Corollary 8.7, f is infinitely renormalizable. The conjecture implies f is a limit of the hyperbolic polynomials g_n; but this contradicts Theorem 4.9, which asserts that c belongs to a non-hyperbolic component of the interior of the Mandelbrot set.

∎

Examples. Every quadratic polynomial with connected Julia set admits z^2 as a quotient.

Some quotients associated to the renormalizable examples of §7.4 are as follows.

I. The map $f(z) = z^2 - 1.772892\ldots$ has a quotient $g_3(z) = z^2 - 1.754878\ldots$, where g_3 is the unique real quadratic polynomial with a critical point of period three.

II. The Feigenbaum polynomial admits quotients $g_2(z) = z^2 - 1$, $g_4(z) = z^2 - 1.310702\ldots, g_{2^n}, \ldots$ of periods $2, 4, 8, \ldots$.

III. For $f(z) = z^2 - 1.54368\ldots$, the map f^2 is simply renormalizable, but since the small postcritical sets $P_2(1)$ and $P_2(2)$ meet, it does *not* admit a quotient of period two.

IV, V. These maps do not admit quotients (other than z^2 and themselves), because their nontrivial renormalizations are crossed (non-simple).

Remarks. The language of quotient maps should help formalize several points in the theory of rational maps. For example, a hyperbolic rational map f with connected Julia set always admits a critically finite quotient, where the attracting cycles are replaced by superattracting cycles. This critically finite map is constructed in [Mc1] and provides a "center" for the component of the space of hyperbolic rational maps containing f. Another potential application is to the "tuning" construction of Douady and Hubbard (see [Dou1], [Dou2], [Mil1] and §7.4). That is, one would like to reconstruct a simply renormalizable mapping f from its quotient g of period n and from the polynomial h to which $f^n : U_n \to V_n$ is hybrid equivalent. (If this can be done, one says f is the tuning of g by h.)

From quotients to renormalization. We will now give a converse to Theorem B.20.

Theorem B.24 (Divisible implies renormalizable) *If the map* $f(z) = z^2 + c$ *admits a proper critically finite quotient* $g(z) = z^2 + c'$, *where* $c' \neq 0$, *then* f^n *is simply renormalizable, where* $1 < n < \infty$ *is the renormalization period of* g.

Recall that the renormalization period of g is the least $n > 1$ such that g^n is simply renormalizable (§8.2). The proof will be based on the lamination criterion for renormalization developed in §8.5. To apply that criterion, we will show that $J(f)$ inherits many of the combinatorial identifications present in the Julia set of g.

It is likely that when $J(f)$ is locally connected, there is a semi-conjugacy $J(g) \to J(f)$, but we will establish somewhat less than this.

For $f(z) = z^2 + c$ a quadratic polynomial with connected Julia set, let

$$\rho_f : (\mathbb{C} - \overline{\Delta}) \to (\mathbb{C} - K(f))$$

denote the Riemann mapping normalized so that $\rho_f(z)/z \to 1$ as $z \to \infty$ (compare §6.2). We let $R_t(f)$ denote the external ray with angle $t \in \mathbb{R}/\mathbb{Z}$; thus $R_t(f) = \rho_f(\exp(2\pi i t)(1, \infty))$.

Theorem B.25 *Let $\phi : f \to g$ be a quotient map between quadratic polynomials, where g is critically finite. Let $s, t \in \mathbb{Q}/\mathbb{Z}$ be a pair of rational external angles such that the external rays $R_s(g)$ and $R_t(g)$ land at a common point in the Julia set of g.*

Then the rays $R_s(f)$ and $R_t(f)$ also land at a common point in the Julia set of f.

In the lamination terminology of §6.4, $\lambda_{\mathbb{Q}}(g) \subset \lambda_{\mathbb{Q}}(f)$.

Proof. If g is *not* a proper quotient of f, then f and g are combinatorially equivalent, and by Theorem B.2 f and g are conformally conjugate. So the theorem is immediate in this case. Also, if $g(z)$ is conjugate to z^2, no rays are identified for g and the theorem is immediate.

Now assume g is a proper quotient of f, and $|P(g)| > 2$. Then the combinatorial quotient map ϕ sends infinity to infinity, since this critical point is distinguished by total invariance. By Theorem B.19, the critical points of g are periodic, so $P(g) \cap J(g) = \emptyset$.

The combinatorial quotient map ϕ is represented by a map ϕ_0 such that

$$\phi_0(z) = \rho_g(\rho_f^{-1}(z))$$

in a neighborhood of infinity (since this condition can always be arranged by isotopy). Equivalently, we may assume ϕ_0 provides a conformal conjugacy between f and g near infinity. By the definition of quotients of branched coverings, there is a lift ϕ_1 of ϕ_0 such that the diagram

$$(\widehat{\mathbb{C}}, P(f)) \xrightarrow{\phi_1} (\widehat{\mathbb{C}}, P(g))$$

$$f \downarrow \qquad\qquad g \downarrow$$

$$(\widehat{\mathbb{C}}, P(g)) \xrightarrow{\phi_0} (\widehat{\mathbb{C}}, P(g))$$

commutes, and ϕ_1 is homotopic to ϕ_0. Thus ϕ_1 also provides a con-formal conjugacy between f and g near infinity. Such a conjugacy is unique, so $\phi_1(z) = \phi_0(z)$ for z large. The homotopy between ϕ_0 and ϕ_1 may be chosen so that $\phi_t(z)$ is constant for z large.

Now suppose $R_s(g)$ and $R_t(g)$ land at a common point x in the Julia set of g for distinct rational angles s and t. Since the angles are rational, the forward orbit E of x is finite. At least two rays land at x, and there are no critical points in the Julia set of g, so at least two rays land at every point in E.

Let λ be the finite lamination corresponding to the rays landing in E, and let Θ be the support of λ. Then (in the notation of §6.4) there is an invariant λ-ray system

$$\sigma : \bigcup_\Theta S_t \to \widehat{\mathbb{C}} - P(g)$$

such that $\sigma(z) = \rho_g(z)$ for $|z| > 1$.

To lift σ to a λ-ray system for f, define

$$\sigma_t(z) = \phi_t^{-1} \circ \sigma(z).$$

This map is well-defined because ϕ_t^{-1} is injective outside of the post-critical set $P(g)$. Then σ_t is a λ-ray system for f, and

$$f(\sigma_1(z)) = f(\phi_1^{-1}(\sigma(z))) = \phi_0^{-1}(g(\sigma(z))) = \phi_0^{-1}(\sigma(z^2)) = \sigma_0(z^2),$$

so σ_1 is a lift of σ_0. The family σ_t provides an isotopy between σ_0 and σ_1, so σ_0 is an f-invariant λ-ray system.

By Theorem 6.14, λ is a subset of the rational lamination of f. In particular, the external rays $R_s(f)$ and $R_t(f)$ land at the same point in the Julia set of f.

∎

Proof of Theorem B.24 (Divisible implies renormalizable).
Since $c' \neq 0$, the critical point $z = 0$ of g is periodic with period
$m > 1$.

Let $\phi : f \to g$ be a quotient map. The sets

$$P_k \;=\; P(f) \cap \phi^{-1}(g^k(0)), \quad k = 1, \ldots, m$$

partition $P(f) \cap \mathbb{C}$ into m disjoint pieces which are cyclically permuted by f. Thus the α fixed point of f is repelling (since otherwise
α belongs to every P_k). Moreover, the forward orbit of the critical
point is disjoint from the α fixed point of f, since $m > 1$.

By the preceding result, the α-lamination of g is contained in the
α-lamination of f. It is easy to see g^m is renormalizable, since g is
expanding on the boundary of the immediate basin of attraction of
$z = 0$. Thus the renormalization period n of g is finite ($1 < n \leq m$).
By Theorem 8.12, f^n is also simply renormalizable.

■

Remarks. The Yoccoz puzzle can be generalized to a version built
from the rays landing at an arbitrary repelling cycle rather than the
α fixed point. Using this version, one may establish the allied result
that f^n is simply renormalizable where n is the period of $z = 0$ for
g.

The construction of the renormalization $f^n : U_n \to V_n$ via the
Yoccoz puzzle is rather ineffective, because it depends on a lineariza-
tion of the α-fixed point to create enlarged puzzle pieces and to obtain a definite annulus between U_n and V_n. It would be very useful
to have a more effective construction depending just on the geometry
of the postcritical set $P(f)$.

Bibliography

[Ah1] L. Ahlfors. *Lectures on Quasiconformal Mappings.* Van Nostrand, 1966.

[Ah2] L. Ahlfors. *Conformal Invariants: Topics in Geometric Function Theory.* McGraw-Hill Book Co., 1973.

[AB] L. Ahlfors and L. Bers. Riemann's mapping theorem for variable metrics. *Annals of Math.* **72**(1960), 385–404.

[BGS] W. Ballman, M. Gromov, and V. Schroeder. *Manifolds of Nonpositive Curvature.* Birkhauser, 1985.

[Bea1] A. Beardon. *The Geometry of Discrete Groups.* Springer-Verlag, 1983.

[Bea2] A. Beardon. *Iteration of Rational Functions.* Springer-Verlag, 1991.

[BP] A. Beardon and C. Pommerenke. The Poincaré metric of plane domains. *J. Lond. Math. Soc.* **18**(1978), 475–483.

[BR] L. Bers and H. L. Royden. Holomorphic families of injections. *Acta Math.* **157**(1986), 259–286.

[Bie] L. Bieberbach. *Conformal Mapping.* Chelsea, 1953.

[Bl] P. Blanchard. Complex analytic dynamics on the Riemann sphere. *Bull. AMS* **11**(1984), 85–141.

[BH] B. Branner and J. H. Hubbard. The iteration of cubic polynomials, Part II: Patterns and parapatterns. *Acta Mathematica* **169**(1992), 229–325.

[Bro] H. Brolin. Invariant sets under iteration of rational functions. *Ark. Math.* **6**(1965), 103–144.

[Bus1] P. Buser. The collar theorem and examples. *Manuscripta Math.* **25**(1978), 349–357.

[Bus2] P. Buser. *Geometry and Spectra of Compact Riemann Surfaces.* Birkhauser Boston, 1992.

[Car1] C. Carathéodory. *Conformal Representation.* Cambridge University Press, 1952.

[Car2] C. Carathéodory. *Theory of Functions of a Complex Variable*, volume II. Chelsea, 1954.

[CG] L. Carleson and T. Gamelin. *Complex Dynamics.* Springer-Verlag, 1993.

[CL] E. F. Collingwood and A. J. Lohwater. *The Theory of Cluster Sets.* Cambridge University Press, 1966.

[Cvi] P. Cvitanović. *Universality in Chaos.* Adam Hilger Ltd, 1984.

[Dou1] A. Douady. Systèmes dynamiques holomorphes. *Astérisque* **105-106**(1983), 39–64.

[Dou2] A. Douady. Chirurgie sur les applications holomorphes. In *Proceedings of the International Conference of Mathematicians*, pages 724–738. American Math. Soc., 1986.

[DH1] A. Douady and J. Hubbard. *Étude dynamique des polynômes complexes.* Pub. Math. d'Orsay, 1984.

[DH2] A. Douady and J. Hubbard. On the dynamics of polynomial-like mappings. *Ann. Sci. Éc. Norm. Sup.* **18**(1985), 287–344.

[DH3] A. Douady and J. Hubbard. A proof of Thurston's topological characterization of rational maps. *Acta Math.* **171**(1993), 263–297.

[EL] E. L. Eremenko and M. Lyubich. The dynamics of analytic transformations. *Leningrad Math. J.* **1**(1990), 563–634.

[Fatou1] P. Fatou. Sur les équations fonctionnelles; (Premier mémoire). *Bull. Sci. Math. France* **47**(1919), 161–271.

[Fatou2] P. Fatou. Sur les équations fonctionnelles; (Deuxième mémoire). *Bull. Sci. Math. France* **48**(1920), 33–94.

[Fatou3] P. Fatou. Sur les équations fonctionnelles; (Troisième mémoire). *Bull. Sci. Math. France* **48**(1920), 208–314.

[Gant] F. R. Gantmacher. *The Theory of Matrices*, volume II. Chelsea, 1959.

[Garn] J. B. Garnett. *Bounded Analytic Functions*. Academic Press, 1981.

[Hub] J. H. Hubbard. Local connectivity of Julia sets and bifurcation loci: three theorems of J.-C. Yoccoz. In L. R. Goldberg and A. V. Phillips, editors, *Topological Methods in Modern Mathematics*, pages 467–511. Publish or Perish, Inc., 1993.

[Julia] G. Julia. Mémoire sur l'itération des applications fonctionnelles. *J. Math. Pures et Appl.* **8**(1918), 47–245.

[Lat] S. Lattès. Sur l'iteration des substitutions rationelles et les fonctions de Poincaré. *C. R. Acad. Sci. Paris* **166**(1918), 26–28.

[LV] O. Lehto and K. J. Virtanen. *Quasiconformal Mappings in the Plane*. Springer-Verlag, 1973.

[Lyu1] M. Lyubich. On typical behavior of the trajectories of a rational mapping of the sphere. *Soviet Math. Dokl.* **27**(1983), 22–25.

[Lyu2] M. Lyubich. An analysis of the stability of the dynamics of rational functions. *Selecta Math. Sov.* **9**(1990), 69–90.

[Lyu3] M. Lyubich. Geometry of quadratic polynomials: Moduli, rigidity, and local connectivity. *Stony Brook IMS Preprint 1993/9.*

[Lyu4] M. Lyubich. On the Lebesgue measure of the Julia set of a quadratic polynomial. *Stony Brook IMS Preprint 1991/10.*

[MSS]   R. Mañé, P. Sad, and D. Sullivan. On the dynamics of rational maps. *Ann. Sci. Éc. Norm. Sup.* **16**(1983), 193–217.

[Mc1]   C. McMullen. Automorphisms of rational maps. In *Holomorphic Functions and Moduli I*, pages 31–60. Springer-Verlag, 1988.

[Mc2]   C. McMullen. Rational maps and Kleinian groups. In *Proceedings of the International Congress of Mathematicians Kyoto 1990*, pages 889–900. Springer-Verlag, 1991.

[Mc3]   C. McMullen. Frontiers in complex dynamics. *Bull. AMS* **31**(1994), 155–172.

[Mc4]   C. McMullen. Renormalization and 3-manifolds which fiber over the circle. *Preprint.*

[McS]   C. McMullen and D. Sullivan. Quasiconformal homeomorphisms and dynamics III: The Teichmüller space of a holomorphic dynamical system. *Preprint.*

[MeSt] W. de Melo and S. van Strien. *One-Dimensional Dynamics.* Springer-Verlag, 1993.

[Mil1] J. Milnor. Self-similarity and hairiness in the Mandelbrot set. In M. C. Tangora, editor, *Computers in Geometry and Topology.* Lect. Notes Pure Appl. Math., Dekker, 1989.

[Mil2] J. Milnor. Dynamics in one complex variable: Introductory lectures. *Stony Brook IMS Preprint 1990/5.*

[Mil3] J. Milnor. Local connectivity of Julia sets: Expository lectures. *Stony Brook IMS Preprint 1992/11.*

[Mon]   P. Montel. *Familles Normales.* Gauthiers-Villars, 1927.

[Nam]   M. Namba. *Branched Coverings and Algebraic Functions.* Longman Scientific and Technical, 1987.

[NvS]   T. Nowicki and S. van Strien. Polynomial maps with a Julia set of positive Lebesgue measure: Fibonacci maps. *Preprint, 1993.*

[Oes]   J. Oesterlé. Démonstration de la conjecture de Bieberbach. *Astérisque* **133-134**(1986), 319–334.

[Rees1] M. Rees. Ergodic rational maps with dense critical point forward orbit. *Ergod. Th. & Dynam. Sys.* **4**(1984), 311–322.

[Rees2] M. Rees. Positive measure sets of ergodic rational maps. *Ann. scient. Éc. Norm. Sup.* **19**(1986), 383–407.

[Roy]   H. L. Royden. *Real Analysis.* The Macmillan Co., 1963.

[SN]   L. Sario and M. Nakai. *Classification Theory of Riemann Surfaces.* Springer-Verlag, 1970.

[Sie]   C. L. Siegel. Iteration of analytic functions. *Annals of Math.* **43**(1942), 607–612.

[Stein] E. M. Stein. *Singular Integrals and Differentiability Properties of Functions.* Princeton Univeristy Press, 1970.

[Str]   K. Strebel. *Quadratic Differentials.* Springer-Verlag, 1984.

[Sul1] D. Sullivan. On the ergodic theory at infinity of an arbitrary discrete group of hyperbolic motions. In *Riemann Surfaces and Related Topics: Proceedings of the 1978 Stony Brook Conference.* Annals of Math. Studies 97, Princeton, 1981.

[Sul2] D. Sullivan. Conformal dynamical systems. In *Geometric Dynamics*, pages 725–752. Springer-Verlag Lecture Notes No. 1007, 1983.

[Sul3] D. Sullivan. Quasiconformal homeomorphisms and dynamics I: Solution of the Fatou-Julia problem on wandering domains. *Annals of Math.* **122**(1985), 401–418.

[Sul4] D. Sullivan. Bounds, quadratic differentials and renormalization conjectures. In F. Browder, editor, *Mathematics into the Twenty-first Century: 1988 Centennial Symposium, August 8-12*, pages 417–466. Amer. Math. Soc., 1992.

[ST] D. Sullivan and W. P. Thurston. Extending holomorphic motions. *Acta Math.* **157**(1986), 243–258.

[Sw] G. Świątek. Hyperbolicity is dense in the real quadratic family. *Stony Brook IMS Preprint 1992/10.*

[Th1] W. P. Thurston. *Geometry and Topology of Three-Manifolds.* Princeton lecture notes, 1979.

[Th2] W. P. Thurston. On the combinatorics and dynamics of iterated rational maps. *Preprint.*

[Yam] A. Yamada. On Marden's universal constant of Fuchsian groups. *Kodai Math. J.* **4**(1981), 266–277.

[Yoc] J.-C. Yoccoz. Sur la connexité locale des ensembles de Julia et du lieu de connexité des polynômes. *In preparation.*

Index

213

9 780691 029818